$$Acsess = c^7 t^5$$

$$Sess = c^6 t^6$$

$$Moen = c^7 t^4$$

$$Moscop = c^6 t^5$$

$$Scopt = c^5 t^6$$

$$Energy = c^6 t^4$$

$$Scop = c^5 t^5$$

$$Moacscat = c^6 t^3$$

$$Momentum = c^5 t^4$$

$$Spatdi = c^4 t^5$$

$$Acscat = c^5 t^3$$

$$Scat = c^4 t^4$$

$$Moacsp = c^5 t^2$$

$$Mosp = c^4 t^3$$

$$Spat = c^3 t^4$$

$$Acsp = c^4 t^2$$

$$Space = c^3 t^3$$

# SPACE PARTICLE THEORY

A Framework for a Theory of Everything

ERIC MITCHELL HORN

Printed in the United States of America

Cover Design by Melissa K. Thomas

First Edition: July 2020

Horn, Eric Mitchell
     Space Particle Theory: A Framework for a Theory of Everyting / Eric Mitchell Horn
     Includes poem; glossary of selected terms and symbols; theory summaries, both text and framework, theory text, prose and mathematical equations, appendices, and references

     1. Space Particle Theory
     2. Framework - Theory of Everything
     3. Physics equations - New to science
     4. Quantum gravity

For further information, please address:
Space Particle Press
P.O. Box 5031
Klamath Falls, OR 97601
www.spaceparticletheory.com

LCCN: 2020914034
ISBN: 978-1-64388-446-2

*Dedication*


*To the wonderful and beautiful women in my life*


*Alice, Terri, Karen, and Kelly*

# Pondering Time

Pondering time; time and time again
Yet to understand time; time and time again
Anyone? Why does an orange squared appear?
Please, please, tell me!
No answer
Please, I beseech you!
Dead
Dead silence
Like a canyon, long lost, filled with dead echoes of time past

Oh to ponder time
To have time to ponder
To ponder time
A wonderful gift
But the time block remains

The necessary criticisms crush
Like the shoe about to crush the ant
Fear not the shoe
Fear the unseen foot

Oh Newton!
Oh Planck!
Oh Einstein!
Rejoice!
What magnificent buildings you have given us!

Destruction on the way
Math doesn't lie
Can it?
Fear the orange squared
Lurking
Lurking, unseen
But there......there......still there
As surely as light travels at the speed of light forever
Forever there

Time, that wonderfully horrendous gift remains
The proteins of the crushed ant remain
The blocks to the story crushed remain
The lurking orange squared remains
Unseen, uncrushed

Life begins anew
Bacteria and fungi grow
The story blocks rearranged
More solid than before
The buildings of Newton, Planck and Einstein
Built anew, on firmer block foundation

Fear the pondering time, the time block remains
Beware the lurking, ...... unseen, ...... uncrushed, ...... orange squared

# CONTENTS

# Future High School Physics Problem

Calculate the gravitational force of attraction between a 1 kg rest-mass and a 2 kg rest-mass 1 meter apart in free-space ignoring temperature and magnetic effects. Show Newton's method and Newton's gravitational constant and the SPT interpretation with the derived SPT gravitational constant and its factors.

By Newton, ($G$ determined experimentally):
$$F_g = G \frac{m_1 m_2}{d^2}$$

$G = 6.6742 \times 10^{-11}$ m³kg⁻¹s⁻²

$$F_g = 6.6742 \times 10^{-11} * \frac{1*2}{1^2} = 1.3348 \times 10^{-10} \text{N}$$

SPT interpretation, ($k_{g''}$ determined theoretically):
$$g'' = k_{g''} * \frac{\varsigma_1 \varsigma_2}{c^2 t_d^2}$$

$$k_{g''} = \frac{g''_{Q_e}}{F''_{Q_e} 4\pi c t_f t_{\varsigma_{Q_e}}^2} = 2.9979 \times 10^8 \text{ m⁻¹s⁻²}$$   (The numerical value of the speed of light in different units.)

$g''_{Q_e} = 1.4044 \times 10^{-132}$ m⁵s⁻²

$F''_{Q_e} = 5.1362 \times 10^{-47}$ m⁵s⁻²

$c = 2.9979 \times 10^8$ ms⁻¹

$t_f = 2.6301 \times 10^{-52}$ s

$t_{\varsigma_{Q_e}} = 9.5943 \times 10^{-27}$ s

$$g'' = \frac{1.4044 \times 10^{-132} * 2.2262 \times 10^{-19} * 4.4524 \times 10^{-19}}{5.1362 \times 10^{-47} * 4 * \pi * 2.9979 \times 10^8 * 2.6301 \times 10^{-52} * (9.5943 \times 10^{-27})^2 * 1^2}$$

$$= 2.9715 \times 10^{-29} \text{m⁵s⁻²} \div 2.2262 \times 10^{-19} = 1.3348 \times 10^{-10} \text{N}$$

# Preface

"Why are you using my iPad?" my wife asks.
"Because I'm rewriting my book."
"Why? What's wrong with your book?"
"It needs to appeal to a larger audience."
"But why my iPad? You're getting your fingerprints all over it."
"That's right. That's what iPads are for. You haven't used this in ten years, so I'm using it now."
"But, I..., I like my pristine iPad. Isn't your iPad working?"
"It's okay. Your iPad will survive of few of my grubby paw prints."

So begins another day of rewriting. Taking over possession of my wife's unused iPad has forced me to start anew. This book should be of interest to the person who has a strong curiosity about how the universe works, to one who wants to understand such things as "dark energy" and "dark matter."  The significant equations  presented should be of interest to the general physics community. Writing a book which appeals to such a wide ranging audience is challenging, much more so than I realized when I first started this project. To get the attention of the university physics professor, I need the equations to "speak truth," equations that can be instantly grasped and recognized as a true advancement in the understanding of how the universe works. Hence the pre-title and post glossary pages, and the Future High School Physics Problem, the Framework Summary and Theory Summary presented initially within the text.

Usually a physics theory is published in physics literature with an audience for a specific physical field of study. This Space Particle Theory (SPT) is being published for an audience of the interested nonprofessional and the general physics community simultaneously. SPT is a foundational theory. As such, it should be of interest to both audiences. Most physics professionals will likely skip the prose and the algebraic steps and primarily be interested in the significant equations, and only then triple check the equations. The prose and algebraic steps are primarily intended for a nonprofessional audience.

To maintain the interest of a nonprofessional audience, I need to relate the equations to everyday life. Hence this preface and the conceptual analogies which are presented within the text. The conceptual analogies, considered in pieces, over many years, were my starting place for the development of SPT, the equations being developed from the conceptual analogies. The concepts of moving volumes, moving volumes through time, moving areas, moving areas through time and moving distances are developed from conceptual analogies involving particles and moving particles. Up to this point in time, I haven't discovered a particle analogy that in some way does not relate to SPT. For this reason, a technical person who works with particle flows should easily be able to relate SPT to their particular field of interest.

It is my firm belief that this theory captures the essence of how the universe works. If I have done my job and presented this theory clearly, and if you have knowledge of basic algebra, including exponents, and a sincere desire to gain an understanding of the basic processes which govern the universe, then this book is for you.

Space Particle Theory (SPT) answers many big questions currently of interest to the general physics community, questions of "dark matter," "dark energy," and why the sun's corona is hotter than the surface of the sun. It also presents a theoretical avenue to quantum gravity. In answering these questions, as in much of science, new questions arise, and new possibilities will be conceived and discovered.

This book is the culmination of a journey, a journey through time and space in a multitude of ways, with a multitude of meanings. It is my intent that, by the end of the book, the reader will understand some of the true possibilities expressed by these words.

Let me start at the beginning, with the beginning of this time scale occurring in 1963 or thereabouts. I was in fifth grade. This was the first time that I can remember receiving an actual science book at the beginning of a school year. As far as I know, prior to this time in my life, I did not have any exposure, in any true sense, to science. I remember looking at the book cover and seeing a depiction of the solar system complete with the sun and the nine planets, Pluto was then classified as a planet. I wish it was, still.

I could not wait to open the book. To my great disappointment, this first lesson in astronomy was toward the end of the book. As I recall, volcanoes and earth science, plants, a little animal biology, and dinosaurs preceded the planets. In 1963 there was no PBS, at least in Oskaloosa, a small town in southern Iowa, where I lived at the time. As a child, I never watched any science programs on television, nor did I read any books relating to science. By the time we discussed the planets and the solar system in class, I had read the chapter months before. A major question remained unanswered even after class discussion. In my mind, I imagine the discussion went something like this:

"Why do the planets go around the sun?"
"Gravity causes the planets to go around the sun," Mrs. Jones replied.
"But, how? What's gravity?"
"Gravity is a force that causes the planets to revolve around the sun."
"What causes gravity?"
"The sun and the planets."
"How do they cause gravity?"
"Be patient, you will learn this in the upper grades! Class! I am handing out a worksheet. Please draw in the planets in their proper orbits, label them, and color them..."

Does this sound familiar? I would imagine this scenario might occur a few grades earlier nowadays, but still essentially with the same result. At the time, I just accepted Mrs. Jones's answer. According to her, I would learn about gravity and force in higher grades. Looking back, how is it that we lose the ability to get to the essence of an issue? We somehow, at sometime in our development, are no longer able to ask the fundamental questions. Is it because we have been told we will learn the answer in a higher grade? Is it because our educational system requires us to regurgitate facts that have been spoon fed to us by our teachers? Do we become numb as a result of this process? Why do we just stop asking the questions *"What is...?"*... *"How does...?"* or *"Why...?"* ? These questions are some of the first scientific questions we ask in our scientific lives. But these questions are often the hardest to answer... even if a true answer is known.

I just know that I wanted an answer in 5th grade; not as a senior in high school physics class, and not as an underclassman in college. By the time I was in college, I accepted that gravity kept the planets from flying off into space. I had long forgotten how to ask the most important questions. Mrs. Jones, why couldn't you answer my question in 1963?

What *is* gravity? I have gone through my entire life not having an answer to this most basic question. I think that the fundamental need to find answers to simple questions is what makes a scientist a scientist. Answering basic questions justifies the hours spent in the library and the laboratory. It is the need to know, the unanswered questions, the curiosity about how the universe works, a thinking person's puzzle, which has captivated me over the years. This leads to a second major question which has remained in my mind until now.

**What *is* charge?** **What is charge?** **What is charge?** What is charge? What is charge? What is charge?
That is the question that I remember asking myself over and over again as I drifted off to sleep on a hot August evening roughly 50 years ago, after my first high school chemistry class with Mr. Dozark, my engaging chemistry teacher. I remember the amazement I felt when he turned off the lights, cut a piece of elemental magnesium and tossed it into a beaker of water. Sparks flew. He did the same with elemental potassium and elemental sodium. Again, light lit the classroom. As I recall, Mr. Dozark, introduced the concept of charge at this time. He never answered the above question to my full satisfaction. He didn't know. Perhaps nobody knew. Was it just a plus and a minus on a piece of paper; a "+" and a "-" in a mathematical formula for balancing a chemical equation? That question has haunted me all these years. Until recently, I thought that I would go to my grave with a poor understanding of what charge actually is. I now have a much greater understanding of what charge might be.

I loved chemistry; chemistry was magic. Chemistry's existence proved that small worlds, nay, small universes existed. Why hadn't I thought of this before? I was dumbstruck by a revelatory brick. I pictured little solar systems complete with little planets and little moons revolving around the atomic nucleus; worlds within worlds within worlds... The imagination of my earlier childhood was reignited. But, yet again, a fundamental question could not be answered.

Junior year in high school is a year of decision. For many of us, at least for my generation, the decisions we make at this young age may largely determine our life's course. I listened to my parents. "You _are_ going to go to college."..."You _must_ choose a career that will earn an income to support a family."..."You _are_ going to have a family; a wife and children, aren't you?!"... "You _need_ to be able to support them, you know." Alas, the little worlds disappeared and I was brought back to reality.

What career path was I going to choose? The answer, as good as any, came by accident early in my senior year. A high school friend, Peter, had just quit his job as a delivery-boy for one of the two pharmacies in my hometown of Independence, Iowa. The two pharmacies were right next door to one another on Main Street, the only business street in this town of 6,000 people. "You _need_ to get a job so you can go to college." I took the job my friend had vacated and the rest is history as they say. I went to pharmacy school which gave me a career so that I could support a family.

Eventually, the imagination inspired by my first encounter with chemistry was crushed by the dogmatic minutia of all the scientific facts and equations pounded into our budding pharmacists' brains. The love of chemistry was nearly lost in the facts of chemistry. However, the idea of little worlds long submerged in my subconscious was still there; unknown, long forgotten, on life support, but not yet flat-lined.

The little worlds life raft resurfaced to my consciousness in 1980 after watching an episode of Carl Sagan's popular television series "Cosmos" on PBS. Gravity became the key to understanding the mechanics of the universe. Understand gravity, understand charge, and understand the universe. That became my passion and remains my passion to this day.

I believe in the process of science and have a great deal of respect for the professional and academic communities that are associated with the many fields of physics, chemistry and engineering. But, I also firmly believe that those who think outside the box continue to play an important role in science, just as they have in the past. This book is written in that spirit.

Within the last ten years, I have developed an antipathy toward watching television programs that present theories of dark energy and dark matter. Before this last decade, I avidly tried to process this information. Over time I began to have strong feelings that dark matter and dark energy which were not understood or explainable by currently known physics, might mean that matter and energy were not well understood either. I



*Eric Mitchell Horn*

questioned whether I was wasting my time watching these speculative science programs. In 2015, following a back injury which limited my activity to "couch-time," I started to develop this Space Particle Theory (SPT) mathematically with algebraic equations.  This book is the result of that effort.

I do not consider any spiritual or religious aspects within the presentation of this theory. I have not considered the emergence of the universe, nor the fate of the universe in this discussion. I have not considered the broad field of parapsychology with its subfields of psychokinesis, clairvoyance, telepathy, precognition, near-death experiences, apparitional phenomena, and synchronicity. This theory does not address ufology, inter-dimensional sentient beings (Bigfoot), inter-dimensional travel, etc. Nor are Ouija boards, astrology, witch-craft, astral projection, channeling, past life regression, alien abductions, seances, reincarnation, or occultism considered.

I also do not consider the huge question of free will. Are our individual fates, our collective fates, our planet's fate, and ultimately the fate of the universe determined simply by individual flows of space particles?  I certainly hope not. Yet, the proof of the existence of an inner-verse offered by this theory, may allow development of significant areas for theoretical and experimental studies in these fields. I leave these topics to others to spend their "time-bank" pondering such things. Researchers in these fields should not be afraid of the mathematics presented in this book. In the equations, I have presented the facts, just the facts, the cold hard facts, as I see them. Many of the hypotheses presented are prose versions of the equations.

I do not address the number of dimensions in the universe. At this point in the development of SPT, it is not that important. The question arises, "Is the number of dimensions in the universe equal to the number of $c$'s in some final equation necessary to describe that physical element?" I believe this is the case, but this is a question open to debate.

This text is written for a readership with mathematical skills at a level of basic algebra, which is a requirement for an understanding of the mathematical, but not the conceptual presentation of the theory. Knowledge of calculus is not necessary. Where is it written that the mathematics of all physics must be difficult? Isn't the mathematics of $F = ma$ easy; likewise, $E = mc^2$ ? I am confident that most readers who have a solid algebraic foundation can understand the math and the concepts presented in this theory, even though the notation of the theory, powers of the speed of light and powers of time, may look frightening at first glance to unfamiliar readers. If you have understood this preface, you can understand this book. If you are a professional, the algebraic steps in this presentation of the theory may be tedious. The algebraic steps are shown for the benefit of the nonprofessional audience. However, the information contained in this theory should be of significant interest to you.

New notation is introduced out of necessity. I encourage the reader to peruse the glossary at the end of this book early, reviewing the notation and its inherent meaning. I urge the reader to refrain from making presumptive judgements; to thoroughly consider the basic concepts, the hypotheses and mathematical foundation of the theory. I also recommend that you physically use your hands, as described in the mathematical introduction. These actions will establish in your mind the concepts of the physical elements of moving distance, moving area, and moving space, which will provide a solid foundation for understanding some of the later concepts presented in this book.

I also encourage the nonprofessional reader to have a pencil and paper available. Several questions are asked throughout the text. Try to answer the questions without looking at the answer that follows. Answering the questions will ensure that you are understanding the presented concepts. The text starts slow and easy, but will become dense very quickly. The text is filled with information and meaning. Take your time to familiarize yourself with the notation, the concepts presented, and their meaning and implications.

Time remains an enigma, powers of time even more-so. I continue to ponder time. Is it possible to spend an eternity of time pondering time? Is our humanly concept of time a physical reality? Perhaps I should read the contemplations of time by others; or would that be a waste of time? Are the mathematics that we use to describe a physical reality capable of truly depicting the physical reality that we experience daily? After all, the physical element of distance, referenced in this theory mathematically by the equation $d = ct$, can be considered to be moving time.

This Space Particle Theory (SPT) has been conceived and developed through a process of analogical conceptualization and thought experiments. The vast majority of the mathematical development of the theory occurred after first conceptualizing a physical phenomena by analogical methods. This is not to say that other conceptualizations than those presented would not be equally valid. Some may criticize this method as not being scientifically pure. Ultimately, it becomes your choice to accept the validity of the analogies presented, choose your own analogies if desired, or choose to consider the mathematical presentation on its own merit. The conceptual analogies have enabled the development of SPT mathematically.

The nonprofessional audience will not initially understand the sections titled "Framework Summary," "Space Particle Theory Summary" or "Technical Introduction" which are directed to professional audiences. If you are a member of the general audience, you will probably want to skip these sections and proceed directly to the "Mathematical Introduction" which is primarily intended for you. Professionals will want exposure to some of the concepts presented in this section as well. The purpose of this book is to describe in relatively simple mathematical terms the processes which govern the universe. Hopefully, by the  end of the book, many nonprofessionals will have some understanding of the initial sections of this book.



*Eric Mitchell Horn*

Although this is a work in progress, I feel the need to publish this theory and share these ideas because of the importance of the subject, and, in my view, because of the resources being expended to pursue theoretical dead ends.  I am confident that the big picture presented by SPT is solid.

Forty years time elapsed between Einstein's publication of $E = mc^2$ and the development of atomic energy. This theory provides a possibility for a revolution in transportation, one that uses electromagnetic energy, as the principal propulsive force in the movement of matter; a propulsive force not based on Newton's third law, i.e. the action-reaction law; but one based on $\dfrac{c^7 t^5}{c^2} = c^5 t^5$ and $\mathbf{H} = c^3 t$.  A light based propulsive system promises to offer the dream of both high efficiency and high speed transportation. Considering the implications of human caused climate change, a massive commitment, similar in scale to the Manhattan or Apollo projects, is needed to develop the technology necessary to accomplish the goal of becoming less dependent on fossil fuels. The development of this technology will simultaneously allow the human species to access the stars in a human lifetime, a prospect enabling humankind to become a true traveller of the cosmos.

# Framework Summary
## Relationships of Key Physical Elements
### Physical Elements by $c$ and $t$ Analysis

This framework summary is presented first in order to generate interest among professionals. Included are the key physical elements hypothesized in this paper, and some other commonly used physical elements. This framework is in an easy to understand format. When moving up and to the right, multiply by $t$ (time). When moving up and to the left multiply by $c$. Note that $c = speed$ where $speed = nc$ or the speed of light if a maximum, depending upon application. $n$ is a time ratio and is incorporated into $t$ for specific applications. For simplicity, the $n$ is dropped in this framework summary. The symbol $\psi$ refers to photonic cycle, and the symbol $\varsigma$ refers to scat, (space multiplied by the speed of light and time.) The symbols $c$ and $t$ are carriers of the units. In the case of SI, these units are the meter and the second.

$c^7t^5$

Acsess, (accelerated sustained energy/space squared)
Moent, (moving energy x time)

$$Acsess = c^7t^5$$
$$ACS_E = Ect$$
$$ACS\varsigma = \varsigma c^3 t$$

$c^6t^6$

Sess, (sustained energy/space squared)

$$Sess = c^6t^6$$
$$S^2 = Et^2$$
$$S^2 = \varsigma c^2 t^2$$
$$\frac{S_{Qe}^2}{\psi} = h't_{R_K}$$

$c^7t^4$

Moacscop

Moen, (moving energy)

$c^6t^5$

Moscop, (moving scop)

Planck's constants, $h'$, $h'_b$

$$h' = \frac{S_{Qe}^2}{t_{R_K}\psi}$$

$c^5t^6$

scopt

$c^6t^4$

Acscop, (accelerated scop)

Energy

$$E = c^6t^4$$
$$E = (\frac{V_S}{t})^2$$
$$1eV = c^6t^4_{k_B}$$
$$E = gct_{g'}$$
$$E = ST$$
$$E = T^2t^2$$

$c^5t^5$

Scop

$$SCP = c^5t^5$$
$$SCP = \varsigma ct$$



*Eric Mitchell Horn*

$$E = \frac{S_{Qe}^2}{t_{R_K} t_\psi}$$

$$E_{P_{max}} = \frac{h'}{t_{R_K}} = \frac{S_{Qe}^2}{t_{R_K}^2} = 4c^2 \varphi_0'^{\,2}$$

$$E_0 = c^2 \varsigma_0 = 1.5764 \times 10^{-163} \text{m}^6/\text{s}^2$$

| $c^6 t^3$ | $c^5 t^4$ | $c^4 t^5$ |
|---|---|---|
| Moacscat | Momentum | Spatdi (spat x distance) |

Power
Radiant intensity $\qquad\qquad p_{max} = c^5 t_\varsigma^4$
Luminous flux
$$P_{max} = c^6 t_\varsigma^3$$

| $c^5 t^3$ | $c^4 t^4$ |
|---|---|
| Acscat (accelerated scat/mass) | Scat/mass |

$$\begin{aligned}
&\text{Force} & \varsigma &= c^4 t_\varsigma^4 \\
&\text{Gravity} & \varsigma &= V_c c\, t_{g'} \\
F_{max} = c^2 V_\varsigma &= c^5 t_\varsigma^3 & \varsigma &= V_\varsigma c\, t_\varsigma \\
g' = c^5 t_V^3 &= c^2 V_c & \varsigma &= (c^3 t_\varsigma^3) c\, t_\varsigma \\
g'' = k_{g''} \frac{\varsigma_1 \varsigma_2}{c^2 t_d^2} & & \varsigma &= (c^3 t_V^3) c\, t_{g'} \\
g'_{obs} = g'_{base}\left(1 - \frac{T_{obs}}{T_P}\right) & & \varsigma_P &= 4\varphi_0'^{\,2} \\
g' = \varsigma \frac{c}{t_{g'}} & & \varsigma_0 &= c^4 t_{R_K}^4 = 1.7540 \times 10^{-180}\text{m}^4
\end{aligned}$$

$$g_0 = c^2 S_0 = 1.1706 \times 10^{-190}\text{m}^5/\text{s}^2$$

| $c^5 t^2$ | $c^4 t^3$ | $c^3 t^4$ |
|---|---|---|
| Moacsp | Mosp, (moving space) | Spat (space x time) |

Mov, (moving volume)
$$mov = c^4 t_V^3$$
$$mov_0 = \frac{\varsigma_0}{t_{g'}} = 3.9048 \times 10^{-199}\text{m}^4/\text{s}$$

Most, (moving space with time)
Charge
$$most = c^4 t_\varsigma^3$$

$$e' = Q_e = c^4 t_{\varsigma Qe}^3 = c^4 \sqrt{t_h^5 t_{R_K}} = (c^2 t_{\varsigma P}^2) c (c\, t_{R_K}) = (A_{\varsigma P})(R_K^{-1}) = A_{\varphi B_{Qe}} Z_0'^{\,-1}$$

$$most_0 = c^4 t_{R_K}^3 = 4.5691 \times 10^{-127}\text{m}^4/\text{s}$$

$$c^4 t^2$$

$V_a$, $S_a$, Acsp, (accelerated space)
Inertial mass, $m_i$
Current

$$S_{a'_{A_\varsigma}} = c^2 A_\varsigma$$

$$S_{a'_{Qe}} = c^4 t^2_{\varsigma Qe}$$
$$acsp_0 = c^4 t^2_{R_K} = 1.1903 \times 10^{-73}\,\text{m}^4/\text{s}^2$$

$$c^3 t^3$$

Space/volume
Entropy
Heat capacity
$$V, V_S, S = c^3 t^3$$
$$V_\varsigma = c^3 t^3_\varsigma$$
$$V_c = c^3 t^3_V$$
$$V_{\varsigma k_B} = k'_B = c^3 t^3_{k_B}$$
$$SE\varsigma = ct\sqrt{\varsigma} = ctA_\varsigma$$
$$SE_E = t\sqrt{E}$$
$$S = t\sqrt{E - \varsigma c^2}$$
$$S_{Qe} = V_{\varsigma Qe} = c^3 t^3_{\varsigma Qe} = c^3 t^2_{\varsigma P} t_{R_K} = \varphi_{B_{Qe}} ct_f$$
$$V_{c_0} = S_0 = \frac{mo\,v_0}{c} = 1.3025 \times 10^{-207}\,\text{m}^3$$

---

$$c^4 t$$

Moacar (moving acar)
Illuminance
Irradiance

Spectral flux density

Heat flux density

$$c^3 t^2$$

Moar (moving area)
Mosa (moving scat-area)
Mova (moving volume-area)
$\varphi_E$, Electric flux

$$\sqrt{E} = \sqrt{h'\nu} = cA_\varsigma = \frac{V_\varsigma}{t_\varsigma}$$

$$\frac{V_s}{t} = c^3 t^2$$

$$c^2 t^3$$

Arti (area x time)

$$\varphi_{E_{Qe}} = c^3 t^2_{\varphi B_{Qe}} = \frac{c^4 t^3_{\varsigma Qe}}{ct_f}$$
$$moar_0 = c^3 t^2_{R_K} = \sqrt{E_0} = 3.9704 \times 10^{-82}\,\text{m}^3/\text{s}$$

---

$$c^3 t$$

$A_a$, Acar, (accelerated area)
Acscar (accelerated scat-area, $S_{a'} = c^3 t_\varsigma$)
$S_{a'}$, Accelerating space

$$c^2 t^2$$

Area
$$A = c^2 t^2$$
Voar (volume-area= $A_V = c^2 t^2_V$)
Scar (scat-area= $A_\varsigma = c^2 t^2_\varsigma$)

Temperature
$H$, Magnetic field strength
Radiosity
Pressure

Newtonian gravitational mass ($A_\varsigma$)
$\varphi_B$, Magnetic flux
Capacitance

$$\varphi_{B_{Qe}} = c^2 t^2_{\varphi B_{Qe}}$$

$$T = \frac{V_S}{t^2}$$

$$\varphi'_0 = c^2 t^2_{\varphi 0} = \frac{c^2 t^3_{\varsigma Qe}}{4\alpha t_f} = \frac{c^2 t^3_{\varsigma Qe}}{2 t_{R_K}} = \frac{c^2 t^2_{\varsigma P}}{2}$$

$$T = c^3 t_\varsigma \qquad\qquad\qquad A_{\varsigma P} = 2\varphi_0'$$

$$T_{loss_g} = c^3 t_\varsigma$$
$$T_0 = c^3 t_{R_K} = 1.0343 \times 10^{-28} \text{m}^3\text{/s}^2$$

$c^3$

Moacdi
(Moving accelerated distance)
Thermal conductivity

$c^2 t$

Modi, (moving distance)
Voltage
Conductance
Electric flux density
$$Z_0'^{-1} = c^2 t_f$$
$$R_K'^{-1} = c^2 t_{R_K} = 2\alpha c^2 t_f$$
$$G_0' = c^2 t_{G_0} = 4\alpha c^2 t_f$$

$c t^2$

Motisq (moving time squared)

$c^2$

Acdi, (accelerated distance)
$A_{a'}$, Accelerating area
$$A_{a'} = n c^2$$
Magnetic reluctance

Specific energy

Absorbed dose

$ct$

$$d = ct$$
Distance/ Moti (moving time)
Permittivity

$$\varepsilon_0' = c t_f = \frac{\lambda t_f}{t_\psi}$$

$c^2 t^{-1}$

moacrat
(moving accelerated ratio)

$c$

Motion / Speed
**E**, Electric field strength

Electric charge density

$$s = nc$$

$$c = \frac{\varepsilon_0'}{t_f} = \frac{\lambda}{t_\psi} = k_{g'} * t_{g'}$$

$t$

Time
$$t_f = 2.6301 \times 10^{-52} \text{s}$$
$$t_V = \left(\frac{V_c}{c^3}\right)^{\frac{1}{3}}$$
$$t_\varsigma = \left(\frac{\varsigma}{c^4}\right)^{\frac{1}{4}}$$
$$t_f = \frac{t_{\varsigma P}^2 t_{R_K}}{t_{\varphi E_{Q_e}}^2} = \frac{t_{\varphi 0}^2 t_{G_0}}{t_{\varphi B_{Q_e}}^2}$$
$$t_{g'} = \frac{t_\varsigma^4}{t_V^3} = 4.4919 \times 10^{18} \text{s}$$
$$t_{R_K} = 2\alpha t_f = 3.8387 \times 10^{-54} \text{s}$$
$$t_{G_0} = 4\alpha t_f = 2t_{R_K} = 7.6774 \times 10^{-54} \text{s}$$
$$t_{k_e} = 4\pi t_f = 3.3050 \times 10^{-51} \text{s}$$
$$t_{\varsigma P} = t_{\varphi 0}\sqrt{2} = 4.7965 \times 10^{-13} \text{s}$$
$$t_{\varphi 0} = t_{k_J} = \frac{t_{\varsigma P}}{\sqrt{2}} = 3.3917 \times 10^{-13} \text{s}$$

$$c\,t^{-1}$$

Acrat (accelerated ratio)

Accelerating distance

$$d_{a'} = \frac{c}{t}$$

$$k_{g'} = \frac{c}{t_{g'}} = \frac{c\,t_V^3}{t_\varsigma^4} = \frac{\varepsilon_0'}{t_f t_{g'}} = \frac{\lambda}{t_\psi t_{g'}}$$

$$k_{g'} = \frac{G_0'}{c\,t_{g'} t_{G_0}} = \frac{c^3 t_{VQe}^3 k_{J'}}{4\alpha t_f t_{\varsigma Qe}}$$

$$k_{g'} = \frac{c^7 t_{VQe}^3 t_{\varsigma Qe}^2}{h' t_{R_K}}$$

$$k_{g'} = 4\pi k_{tfg} c^2 k_{e''}$$

$$k_{g'} = k_{tfg} c^4 \mu_0'$$

$$k_{g'} = 2\alpha k_{tfg} c^3 R_K'$$

$$k_{g'} = k_{tfg} c^3 Z_0'$$

$$1(n)$$

ratio/number

Magnetic flux density, **B**

$$k_{tfg} = \frac{t_f}{t_{g'}}$$

$$N_A = \left(\frac{t_R}{t_{k_B}}\right)^3$$

$$\alpha = \frac{t_{R_K}}{2t_f} = \frac{t_{\varsigma Qe}^3}{2t_{\varsigma P}^2 t_f}$$

$$k_{tkg\varsigma} = \frac{t_{1kg}^4}{t_{1mtr}^4} = 2.2262 \times 10^{-19}$$

$$c\,t^{-2}$$

mointisq

$$t^{-1}$$

Inti (inverse time)

Frequency

Exposure

$$c^{-1}$$

Inmo (inverse motion)

$$k_{e''} t_{k_e} = \varepsilon_0' Z_0' = R_K' c\, t_{R_K} = \frac{1}{c}$$

$$t^{-2}$$

intisq

$$c^{-1}t^{-1}$$

Indi (inverse distance)

Specific heat capacity

Specific entropy

$$k_{e''} = \frac{1}{c\,t_{k_e}} = \frac{1}{4\pi c\,t_f} = \frac{\varphi_{B_{Qe}}}{4\pi V_{\varsigma Qe}}$$

$$c^{-2}$$

Inacdi (inverse acdi)

Inductance

$$t^{-3}$$

Inticu

$$c^{-1}t^{-2}$$

Inmointisq

$$\frac{1}{2\pi c\,t_{\varsigma P}^2}$$

$$c^{-2}t^{-1}$$

Inmodi

Resistance

$$Z_0' = \frac{1}{c^2 t_f} = 2\alpha R_K'$$

$$R_K' = \frac{1}{c^2 t_{R_K}} = \frac{1}{2\alpha c^2 t_f}$$

$$G_0'^{-1} = \frac{1}{c^2 t_{G_0}} = \frac{1}{4\alpha c^2 t_f}$$

$$c^{-1}t^{-3}$$

Inmointicu

$$k_{g''} = \frac{g''_{Q_e}}{F''_{Q_e} c\, t_{k_e} t^2_{\varsigma Q_e}} = 2.9979 \times 10^8 \,\text{m}^{-1}\text{s}^{-2}$$

$$c^{-2}t^{-2}$$

Inar (inverse area)

$$k_{J'} = \frac{4\alpha t_f}{c^2 t^3_{\varsigma Q_e}} = \frac{2}{c^2 t^2_{\varsigma P}}$$

$$c^{-3}t^{-1}$$

Inacar

Permeability

$$\mu'_0 = \frac{1}{c^3 t_f}$$

# Space Particle Theory Summary

Space Particle Theory (SPT) is a foundational theory. The theory is presented on a level such that individuals who have a strong foundation in algebra and a strong desire to understand how the universe works can understand its many implications.

Conceptual analogies which give rise to major hypotheses of the theory are presented initially in prose format. These hypotheses are subsequently developed and expanded upon in algebraic equations involving physical elements. Most of the theory is presented in a mathematical format, a description in mathematical terms and equations of the basic processes governing the observable universe. Some of these relationships are known. Many are introduced for the first time in this text. This summary details a few of the major results derivable from these major hypotheses. SPT provides a framework for further research and consideration of presently unknown physical elements.

A number of the hypotheses consider the physical element of space and its relationship with other known physical elements. A physical element is defined to be a building block, described mathematically as an equation involving  powers of the speed of light and powers of time, which can be used in combination to create more complex physical elements. A few examples of physical elements include acceleration, speed, distance, area, and volume. These can be used to mathematically describe mass, energy, power, force, charge, etc. which are themselves physical elements or subsets of physical elements.

A foundational hypothesis that space exists as a particle, and that matter is formed from the flow of space particles through a combination of hole-complexes, is proposed. Space particles have no mass and pass at the speed of light between an unobservable part of the universe (inner-verse) and the observable part of the universe (outer-verse) and vice versa. The space particle is proposed to be the only true particle, the particle of exchange in the universe. Detectable massless particles are proposed to be hole-complexes through which space particles pass from the unobservable part of the universe to the observable part of the universe.

The named physical element of **ac**celerated **s**ustained **e**nergy **s**pace **s**quared (acronym acsess) is presented:

$$acsess = mc^3t = Ect$$

Acsess reflects the fact that energy travels at the speed of light through time. Mass and energy are proposed to be opposite processes.

The concept of the physical element of space multiplied by the speed of light and time, (a symbolized acronym of scat, $\varsigma$) is presented whereby scat, not mass as it is currently defined, quantifies matter. Under conditions of slow speed and low gravity:

$$mass \approx matter = restmass = scat(\varsigma) = V_c c t_{g'}$$

The theory ties mass to the speed of light using a consumptive volume of space, $V_c$, traveling at the speed of light through a specified time, $t_{g'}$. A quantity of rest mass is expressed in m⁴, symbolized $m_{m^4}$, and set to equal a scat quantity, $c^4 t_\varsigma^4$, such that:

$$m_{m^4} = V_c c t_{g'} = V_\varsigma c t_\varsigma = A_\varsigma^2 = c^4 t_\varsigma^4 = (m_{kg} * 2.2262 \times 10^{-19})$$

where $V_\varsigma$ is a volume of space consumed by an object with mass in one scat-time, $t_\varsigma$:

$$t_\varsigma = (\frac{m_{m^4}}{c^4})^{\frac{1}{4}}$$

$$V_\varsigma = c^3 t_\varsigma^3$$

The volume-time is given by the equation:

$$t_V = (\frac{V_c}{c^3})^{\frac{1}{3}}.$$

Employing these equations, the well known relationships of $E = mc^2$, $E = Fd$, $F = ma$, $E = QV$, $Q = It$, and many others are easily derived.

A scat-area, $A_\varsigma$, associated with an object with mass is defined:

$$A_\varsigma = c^2 t_\varsigma^2.$$

A time constant, $t_{g'}$, which may be the age of the universe, is a determinant of the derived gravitational constant for one object with mass:

$$k_{g'} = \frac{c}{t_{g'}} = \frac{c t_V^3}{t_\varsigma^4} = 6.674 \times 10^{-11} \text{ m/s}^2.$$

$t_{g'}$ is calculated to be $4.4919 \times 10^{18}$ s.

Several new and significant relationships between the physical elements of energy, scat/mass, gravity, space, charge and temperature are proposed. One outcome of the theory relates energy and gravity with the speed of light by the equation:

$$E = g' c t_{g'}$$

where $g'$ is the gravitational force associated with one object with mass expressed as:

$$g' = \varsigma \frac{c}{t_{g'}}$$

Gravity is related to the consumptive volume of one object with mass by the equation:

$$g'_{base} = g' = c^2 V_c$$

The equations for gravity are used to calculate an instantaneous consumptive volume, $V_c$, and a volume of space consumed per time:

$$\sqrt{E} = \frac{V_\varsigma}{t_\varsigma} = c A_\varsigma$$

Temperature is proposed to be an accelerating volume of space. The temperature loss due to gravity is proposed:

$$T_{loss_g} = c^3 t_\varsigma$$

The observed gravity relationship with temperature is proposed:

$$g_{obs} = g'_{base}\left(1 - \frac{T}{T_P}\right) \quad \text{where } T_P \text{ is Planck temperature (as derived in this text)}$$

A fundamental time constant, $t_f$, is found to be a common determinant in many quantum constants:

$$t_f = 2.6301 \times 10^{-52}\text{s}$$

$$\mu'_0 = \frac{1}{c^3 t_f}$$

$$Z_0' = \frac{1}{c^2 t_f}$$

$$k_{e''} = \frac{1}{4\pi c t_f}$$

$$\varepsilon_0' = c t_f = 7.8848 \times 10^{-44} \text{m}$$

The time-length of the von Klitzing constant is calculated:

$$R_K' = \frac{1}{c^2 t_{R_K}}$$

$$t_{R_K} = 3.8387 \times 10^{-54} \text{s}$$

$t_{R_K}$ relates Planck's constant to the electron charge space squared, $S_{Q_e}^2$:

$$h' * t_{R_K} = S_{Q_e}^2 = c^6 t_{\varsigma_{Q_e}}^6$$

The fine structure constant, $\alpha$, is determined to be a ratio of time-lengths:

$$\alpha = \frac{t_{R_K}}{2t_f} = \frac{t_{G_0}}{4t_f} = \frac{2\pi t_{R_K}}{t_{k_e}}$$

The quantum constants are shown to be related to $k_{g'}$ via combinations of powers of the speed of light, and various combinations of $t_f$, $t_{g'}$ and $t_{R_K}$.

Seven variations of Newton's equation relating the force of attraction between two objects with mass expressed in m⁴ are presented including:

$$g'' = k_{g''} * \frac{\varsigma_1 \varsigma_2}{c^2 t_d^2}$$

where the numerical value of the gravitational attraction constant $k_{g''}$ is the numerical value of the speed of light and has units m⁻¹s⁻²:

$$k_{g''} = 2.9979 \times 10^8 \text{ m}^{-1}\text{s}^{-2}$$

The factors of $k_{g''}$ are determined:

$$k_{g''} = \frac{g''_{Q_e}}{F''_{Q_e} 4\pi c t_f t^2_{\varsigma Q_e}} = 2.9979 \times 10^8 \text{m}^{-1}\text{s}^{-2}$$

The masses in Newton's gravitational equation are found to relate the scat-area of the objects with mass, $A_\varsigma$. This area equals the square root of the scat/mass  represented by Newton's $F = ma$ and Einstein's $E = mc^2$. Physically, the scat-area is considered to be the consumptive area of the rest mass.

Space is related to temperature by the equation:

$$S = Tt^2$$

Energy is related to space and temperature by the equations:

$$E = ST$$

$$S^2 = Et^2$$

$$E = T^2t^2$$

An equation is also presented directly relating the amount of space created following a fusion reaction to energy and mass. The space equivalent of energy represents the space returning to the outer-verse due to energy, and the space equivalent of mass represents the space returned to the inner-verse by an object with mass. The two are related through the relationship:

$$|SE_E| = |t\sqrt{E}| = |ct\sqrt{\varsigma}| = |SE_\varsigma|$$

The theory proposes, with supportive calculations, that energy and mass are related via Einstein's equation $E = mc^2$ interpreted as:

$$\left(\frac{V_S}{t}\right)^2 = (cA_\varsigma)^2$$

$$V_S = t\sqrt{h'\nu}$$

The energy of Einstein's $mc^2$, $E_{oo}$, is proposed to represent the energy leaving the outer-verse.  When mass is changed to energy, the proposed hole-complexes which connect the inner-verse to the outer-verse are proposed to "flip" and the flow of space particles between the two verses is reversed.

The charge of the electron is a very small component of the electron as a whole. The ratio of the rest mass of the electron to the rest mass equivalent of the charge of the electron:

$$\frac{m_e}{m_{Q_e}} = \frac{9.1094 \times 10^{-31}}{3.0744 \times 10^{-52}} = \frac{t_{\varsigma e}^4}{t_{\varsigma Q_e}^4} = \frac{2.5106 \times 10^{-83}}{8.4733 \times 10^{-105}} = 2.9630 \times 10^{21} \ \text{(unit-less)}$$

Electric charge is demonstrated to be a volume of space traveling the speed of light, or an area traveling at the speed of light a very short distance. With SI derived units m⁴s⁻¹, charge relates to energy by the relationships:

$$\frac{Q_e}{ct_{\varsigma Q_e}} = \frac{V_{\varsigma Q_e}}{t_{\varsigma Q_e}} = \sqrt{E_{Q_e}}$$

$$\text{or} \quad e' = Q_e = ct_{\varsigma Q_e}\sqrt{E_{Q_e}}$$

$$= c^4 t_{\varsigma Q_e}^3 = cV_{\varsigma Q_e} = c(c^2 t_{\varsigma P}^2)(ct_{R_K}) = (A_{\varsigma P})(R_K^{-1}) = \varphi_{B_{Q_e}} c^2 t_f = \varphi_{B_{Q_e}} Z_0^{-1}.$$

The numerical value of the charge of the electron, $e'$ or $Q_e$, is calculated:

$$e' = Q_e = 7.1338 \times 10^{-45} \ \text{m⁴/s.}$$

Three variations of Coulomb's law are presented, including:

$$F_Q'' = k_{e''} * \frac{Q_1 Q_2}{(ct_d)^2} = \frac{Q_1 Q_2}{c^3 t^3}$$

$$k_{e''} = 1.0093 \times 10^{42} \ \text{m⁻¹} = \frac{1}{4\pi ct_f} = \frac{\varphi_{B_{Q_e}}}{4\pi V_{\varsigma Q_e}}$$

$$\varphi_{B_{Q_e}} = c^2 t_{\varphi B_{Q_e}}^2$$

A magnetic field strength quantum constant, the inverse of the magnetic constant is hypothesized, where $H_0$ is a directionalized accelerating volume of space (directionalized temperature.)

$$H_0 = \frac{1}{\mu_0} = c^3 t_f = \frac{V_{\varsigma Q_e}}{t^2_{\varphi_{B_{Q_e}}}}$$

Planck's constant is proposed to relate a volume of space returning to the observable universe for each individual photonic cycle. The amount of space returned to the the observable universe represented by Planck's constant is found to be the equivalent amount of space that the charge of the electron consumes in an electron charge scat-time:

$$c\, t_{R_K} \sqrt{\frac{h'}{t_{R_K}}} = c^4 t^3_{\varsigma Q_e}$$

The SPT extension of the Planck-Einstein relation is determined:

$$E = h\nu = \frac{S^2_{Q_e}}{t_{R_K} t_\psi}$$

This supports the hypothesis that $E = (\frac{V_S}{t})^2$

The converted Planck constant:

$$h' = 1.4751 \times 10^{-52} \text{ m}^6/\text{s·cycle}$$

Each photonic cycle returns $2.3796 \times 10^{-53}$ m³ to the outer-verse.

The electron charge gravitational force is calculated:

$$g_{Q_e} = c^5 t^3_{V_{Q_e}} = c^2 V_{c_{Q_e}} = 4.5681 \times 10^{-81} \text{m}^5/\text{s}^2 = 2.0601 \times 10^{-62} \text{N}$$

The quantum electromagnetic force is calculated:

$$F_{Q_e} = c Q_e = c^5 t^3_{\varsigma Q_e} = c^5 t^2_{\varphi Q_e} t_f = c^5 t^2_{\varsigma P} t_{R_K} = 2.1387 \times 10^{-36} \text{m}^5/\text{s}^2$$

$$= 9.6068 \times 10^{-18}\text{N}$$

The quantum strong force is calculated:

$$F_{strong} = 2\frac{t_f}{t_{R_K}} * cQ_e = \alpha cQ_e = 2c^5 t_{\varsigma P}^2 t_f$$

The quantum spin of fermions is calculated:

$$spin_f = \frac{S_{Q_e}^2}{4\pi t_{R_K}} = 1.1739 \times 10^{-53}\text{m}^6/\text{s} = 7.3758 \times 10^{-53}\text{m}^6/\text{s}\cdot\psi$$

The temperature of space, $T_S$, is proposed to be a difference between energy entering the observable universe, $E_{io}$, and energy exiting the observable universe, $E_{oo}$, per volume of space, $V_S$:

$$T_S = \frac{E_{io} - E_{oo}}{c^3 t^3} = \frac{E_S}{V_S}$$

Physically, temperature is an accelerating volume of space. The temperature equivalent of 1 K is calculated to be $0.19438$ m³/s². Given the average temperature of cosmic space equals 2.275 °K, the average cosmic energy of space is calculated:

$$0.4422 \text{ m}^6/\text{m}^3\cdot\text{s}^2 = 1.986 \times 10^{18} \text{ J/m}^3.$$

A gain or loss of entropy is proposed to be a volume of space added to or subtracted from a system:

$$k_B' = c^3 t_{k_B}^3 = 1.5812 \times 10^{-41}\text{m}^3$$

$$t_{k_B} = 8.3722 \times 10^{-23}\text{s}$$ equals the scat-time of one electron volt:

$$1 \text{ eV} = c^6 t_{k_B}^4 = 3.5669 \times 10^{-38}\text{m}^6/\text{s}^2$$

Given the temperature equivalent of 1eV equals $11{,}605$ K, the temperature equivalent of $k_B'$ is calculated:

$$T_{k'_B} = c^3 t_{k_B} = 2.2558 \times 10^3 \text{ m}^3/\text{s}^2$$

Given the Standard Model prediction of the graviton being a spin 2 particle and the requirement that the space particle is space filling, the shape of the space particle could be an acute golden rhombohedron or a rhombic dodecahedron. The formula for volume of an acute golden rhombohedron, $V_{agr} = S_0 = \dfrac{1}{5}\sqrt{10 + 2\sqrt{5}}\, a^3$, is used to determine the length of the side of the space particle, $a$, and the shortest meaningful volume time-length, $t_{V_0}$.

The inverse von Klitzing constant is used to calculate the proposed quantum constants of quantum energy, $E_0$; quantum gravity, $g_0$; quantum scat, $\varsigma_0$; quantum moving volume, $mov_0$; the volume of the space particle, $S_0$; and the quantum temperature gain or loss by one space particle entering or leaving the outer-verse, $T_0$.

$$R_K = \frac{1}{c^2 t_{R_K}} = 2.8985 \times 10^{36} \text{s/m}^2$$

$$R_K^{-1} = 3.4501 \times 10^{-37} \text{m}^2/\text{s}$$

$$(R_K^{-1})^2 = c^4 t_{R_K}^2 = 1.1903 \times 10^{-73} \text{m}^4/\text{s}^2$$

$$\varsigma_0 = \varsigma_{R_K} = (R_K^{-1})^2 * t_{R_K}^2 = c^4 t_{R_K}^4 = 1.7540 \times 10^{-180} \text{m}^4 = 7.8788 \times 10^{-162} \text{kg}$$

$$E_0 = E_{R_K} = c^2 \varsigma_0 = c^6 t_{R_K}^4 = 1.5764 \times 10^{-163} \text{m}^6/\text{s}^2 = 7.0810 \times 10^{-145} \text{J}$$

$$mov_0 = mov_{R_K} = \frac{\varsigma_{R_K}}{t'_g} = 3.8847 \times 10^{-199} \text{m}^4/\text{s}$$

$$g_0 = c * mov_{R_K} = \frac{c\varsigma_0}{t_g} = c^2 S_0 = 1.1646 \times 10^{-190} \text{m}^5/\text{s}^2 = 5.2313 \times 10^{-172} \text{N}$$

$$V_{c_0} = S_0 = \frac{mov_{R_K}}{c} = 1.2958 \times 10^{-207} \text{m}^3$$

$$T_0 = c^3 t_{R_K} = 1.0343 \times 10^{-28} \text{ m}^3/\text{s}^2 = 5.3210 \times 10^{-28} \text{K}$$

$$a_0 = \left(\frac{5S_0}{\sqrt{10 + 2\sqrt{5}}}\right)^{\frac{1}{3}} = (1.7031 \times 10^{-207})^{\frac{1}{3}} = 1.1942 \times 10^{-69}\text{m}$$

$$t_{V_0} = t_a = \frac{a_0}{c} = 3.9834 \times 10^{-78}\text{s}$$

$\varsigma_0$ may be the rest mass of the neutrino.

Although this work remains unfinished at the time of this publication, the $G_F$ factor in the derived Fermi coupling constant is shown to have gravitational, magnetic and $S_0$ ($S_0 = V_{c_{R_K}}$) factors:

$$G_F = \frac{(9.3312 \times 10^{-23})c^3 V_{c_P}^3 V_{c_{H_0}} V_{c_{R_K}} t_{g'}^3}{V_{c_{k_B}}^2 V_{c_{k_e}} t_{g'}^2 \psi^3}$$

This expression has the general form:

$$\frac{c}{t_{g'}}(c^2 V_c^2 t_{g'}^2) = \frac{c}{t_{g'}} \varsigma^2$$

This theory can account for observations that lead to proposals for the existence of dark energy and dark matter and the observation that the sun's corona is hotter than the sun's surface. NASA's EM drive and other possible anti-gravity devices may also be explained by this theory.

# Technical Introduction

Several theories have been proposed for modeling the fundamental nature of the universe. Among the most successful are M-theory which unifies many different versions of superstring theory utilizing S and T duality transformations and eleven dimensional supergravity.[1] However, problems with current theories involve observations that lead to the proposals for the existence of dark matter [2] and dark energy [3], entities that are currently not well understood, as well as difficulties in understanding the observation that the corona of the sun is hotter than the surface of the sun [4,5], and the physics of Nasa's EM drive.[6]

SPT proposes to provide a possible physical explanation for these observations. A model of the universe is presented based on the three observed physical dimensions and several unobserved physical dimensions sharing a common dimension of time, and the observed physical constant of the speed of light. Utilizing simple arithmetic relationships; space, energy, mass, gravity, charge and temperature are related via powers of the speed of light and powers of time.

Some theories suggest that the force of gravity and the other known forces are fundamentally different and arise from fundamentally different mechanisms, while a posit of M-Theory is that all of the known fundamental forces arise from the same fundamental mechanism. Space Particle Theory (SPT) offers a different perspective altogether. This introduction will highlight some of the differences between M-theory and SPT. Theorists involved with loop quantum gravity, and others, can make comparisons with M-theory and SPT and reach their own conclusions.

SPT agrees with a principal tenet of M-theory in that the mechanism resulting in the recognized force of gravity is fundamentally connected to the other known forces, e.g. the strong force, the weak force, and the electromagnetic force. SPT demonstrates that the linkage between force and the charge of the electron is the speed of light. The difference between the gravity of the charge of the electron and the force of the charge of the electron is a time element. SPT reveals that; as an analogy, the gravity of the electron charge relates to an instantaneous photograph of a process, and the force associated with the charge of the electron relates to a movie of that process. In effect, gravity is a fundamental process and many of the other known forces reflect the addition of time elements hidden in the process. SPT considers that gravity is the result of a consumptive side of a process and energy is the result of a supply side of a process.

Whereas M-theory aims to unify Einstein's relativity version of gravity with the other forces, SPT hypothesizes that the quantity of mass is a separate entity from the property of mass; and that Einstein's relativity theory pertains to the property of mass which is one of the factors making the mathematics of the gravity in Einstein's relativity so complex. Instead, SPT reveals a linkage of observed gravity with temperature.



*Eric Mitchell Horn*

SPT hypothesizes that gravity results from the flow of space particles toward a hole-complex which connects an observable outer-verse to an unobservable inner-verse, and that energy is a reverse process, i.e. energy is a flow of space particles out of a hole or hole-complex. Gravity and force are both shown by SPT to be a "moving, moving volume of space," with force having an additional time element.

The proposed existence of an inner-verse at this point of SPT's development is a "black box." The exact structure and relationships between the verses is unknown. Multiple inner-verses probably exist. It may be that, while humans can only experience three physical dimensions at any instant in time, we "transform" between the verses extremely quickly. The time for this transformation may be $t_f$. In fact, we may be creatures inhabiting "movie frames" at a given instant in time as we transform from one verse to another. Much more theoretical work is needed in this area.

SPT considers the wave-particle view of matter to be essentially correct, with a view of matter existing as a hole-complex with the only true particle being the space particle. The individual holes in space can be propagated as a wave, but space itself is a moving collection of space particles. The particle nature of matter directly results from space particles and space particle flows into and out of the individual holes within the hole-complexes which result in the formation of matter. The exact arrangement of the individual holes within the hole-complex at any given instant in time is probably unknowable. At this point in SPT's development, the hole-complexes are another "black box" with a great deal more theoretical work required.

M-theory has no prospect on the horizon for experimental support and many physicists doubt that it reflects any true reality which we experience in day to day interactions with the universe. The linkage between gravity and temperature revealed by SPT is verifiable by experiment.

The Standard Model portrays a set of rules governing the interactions between various particles. SPT hypothesizes that everything which is detectable, is in actual physical reality, a hole or hole-complex or the result of an interaction between hole-complexes. Hole complexes connect an unobservable inner-verse to the observable outer-verse. The only true particle is the space particle, and space particles pass through hole complexes from the inner-verse to the outer-verse and vice versa. SPT posits that the determining factor between energy and matter is the direction of space particle flow between the two verses, and that the speed of the space particle flows the instant before passage between the two verses is the speed of light. Many of the quantum constants in the Standard Model are shown to involve a very few time constants. More theoretical work is needed to include all quantum constants.

The various string theories, which are unified under M-Theory, describe the different sub-particles mathematically as one dimensional strings; i.e. possessing length only, with no width or height. These one dimensional strings vibrate on a brane to form the various entities and interactions observed in the universe. SPT posits that these sub-particles are holes or hole-complexes and that fundamentally space is a quantifiable substance, existing as the only true particle.

Supergravity places a limit on M-theory having eleven dimensions, 7 of which are currently non-detectable. These dimensions are extremely small; i.e. lost to our current observable universe. Within the context of SPT, at this point in its development, the consideration of an inner-verse to an inner-verse and the exact relationships between the verses are deferred. If space, and thus space particles are conserved, then an inner-verse to an inner-verse may be necessary. It may be that three sub-verses which have three physical dimensions, each sharing a common time dimension at the boundaries, create an eleven dimensional universe. A common element between M-theory and SPT may be the existence of an extremely small, undetectable, part of the universe. The number of dimensions in SPT appears to be clearly indicated by the number of $c$'s necessary to describe a physical element. The observation that energy travels at the speed of light requires at least a seven dimensional universe, while the observation that energy can move a mass a distance requires an eleven dimensional universe.

While M-theory involves one dimensional strings, SPT holds as a tenet that the shape of the space particle is discernible via quantum constant relationships. SPT employs a three dimensional space particle, the volume of the space particle calculated using the inverse of the von Klitzing constant, $R_K^{-1}$. The Standard Model predicts that the carrier of gravitational force is a spin 2 particle. Given this requirement and the hypothesis that the space particle must be space filling, an acute golden rhombohedron or a rhombic dodecahedron are proposed to be possible shapes of the space particle.

M-theory incorporates five different string theories with the inclusion of a membrane and S and T duality transforms. SPT requires a boundary between the observable and the unobservable universe. In this sense, the boundary of SPT may reflect the membrane involved in M-theory.

Many physicists are attracted to M-theory because of its simplicity over other possible theories. SPT, which requires only a basic understanding of algebra, is even simpler. It offers the real possibility for an introductory quantum mechanics course taught in high school.

M-theorists are currently involved in efforts to "compactify" extra dimensions. SPT holds that these dimensions are unobservable and at this stage of development, the process of "compactification" is not essential to the theory.

Whereas M-theory provides a connection to Einstein's relativity, SPT holds that no direct connection between gravity and relativity is necessary. SPT hypothesizes that the quantity of mass of Newton's second law, $F = ma$, which is the same as the mass considered in Einstein's $E = mc^2$, is different than the property of mass considered in Einstein's relativity. It is also shown that these "masses" are both different from the mass of Newton's gravitational equation. Essentially $m_{E_R} \neq m_{N_g} \neq m_{N_{2ndlaw}}$. For this reason, a new symbol, $\varsigma$ (termed "scat"), symbolizing a quantity of space multiplied by the speed of light and time, is introduced.

With M-theory, one of the vibrational states of a zero dimension point particle of a string gives rise to a graviton. A graviton is a quantum mechanical particle that "carries" the gravitational force. SPT considers the proposed graviton and the space particle to be one and the same, except the graviton or space particle, symbolized $S_0$, is three dimensional. SPT holds that there is no such thing as a zero dimension point particle which leads to the hypothesis that there is no such thing as empty space.

# Mathematical Introduction and Theory Foundation

Note to nonprofessionals: Take time to understand this section thoroughly. It is the foundation, the bedrock, of SPT.

Note to professionals: Specific numerical values are understood to be approximations, i.e. = means ≈ ... Because of rounding errors which are multiplied when carrying specific values to powers, the five numerals usually used in representing a specific value may be accurate to only three significant digits. Please excuse this discretion. Values for the quantum constants are NIST [National Institute of Standards and Technology] values with the exception of the gravitational constant. Professionals may wish to first consider the summary of foundational hypotheses presented in Appendix F.

*Definition:* A **hypothesis** is a supposition, conjecture, or proposed explanation made on the basis of limited evidence, which is used as a starting point for further investigation.

*Definition:* A **physical element** is a building block for the structure of this Space Particle Theory (SPT). Physical elements may be multiplied together or divided to form other physical elements. Physical elements are related by physical constants.

*Definition:* A **physical constant** specifies a physical quantity which relates two physical elements. A physical constant will always have a numerical value associated with it. A physical constant is itself a subset of a physical element. The speed of light symbolized $c$, is an example of a physical constant. It has an approximate value in Systeme International (SI) units of $2.9979 \times 10^8$ m/s. The speed of light is itself a subset of the physical element of motion or speed. Physical constants are thought (?) to be invariable with time.

*Definition:* A **time-length** is defined to be a length of time, a time interval, or a period of time, symbolized $t$ for the physical element of time in general; or $t$ with a right-sided subscript for a particular subset of the physical element of time as in $t_\varsigma$ or $t_V$; or a specific time-length as in $t_{R_K}$ which may be associated with a particular physical constant. These specific notations for time are defined throughout the paper and in the glossary.

*Definition:* A **Systeme International (SI) unit**, is an internationally recognized unit of measure within the metric system. Each physical element will have an associated derived SI unit of measure, except for the physical element of ratio/number which is a unit-less ratio of the same physical element or combination of physical elements to yield a unit-less number.

*Definition:* A **derived SI unit** is a unit of measure involving only powers of the meter and powers of the second. An example is the derived SI unit of a meter to the 4th power per second squared, symbolized m⁴/s². Current will be shown to have a SI derived unit of a m⁴/s².

*Definition:* A $ct$ **expression** is a mathematical representation of a physical element involving powers of the speed of light and powers of time. For example, the $ct$ expression for the physical element of area is $c^2t^2$.

*Definition:* A $ct$ **equation** is a mathematical equation involving powers of the speed of light and powers of time which describes a physical element. For example, the $ct$ equation for the physical element of area is $A = c^2t^2$, and a $ct$ equation for a specific area can be given by $A = c^2t_d^2$ where $t_d$ is a specific time-length.

**Hypothesis:** *All physical elements and their subsets; e.g. time, distance, moving distance, area, moving area, volume/space, moving volume/space, scat/mass, energy, gravity, space, temperature, charge, etc. can be expressed in terms of powers of the speed of light and powers of time.* These $ct$ expressions, derived throughout this book, represent physical elements.

**Hypothesis:** *All physical constants can be expressed as ratios between various combinations of powers of the speed of light, and powers of time, with a specific time-length given for a specific constant.*

With the above definitions and hypotheses in mind:

Buckle up. We are in for a ride. You are the driver and I am the passenger. Imagine that we are in a Model T traveling at 20 miles per hour.

*Question:* If we drive for one hour, how many miles have we traveled?
Answer:   20 miles per hour X 1 hour equals 20 miles or $20 * 1 = 20$ miles

*Question:* What two factors determine the distance we travel?
Answer:  Speed and time. Speed of travel multiplied by time of travel equals distance of travel:

$$speed \times time = distance$$

Consider the speed of light, which physicists symbolize with the italicized letter $c$. The speed of light is extremely fast compared to the speeds that we encounter in our everyday lives. It is roughly 186,000 miles per second, or in SI units, 2.9979 x 10⁸ meters per second, (m/s).

*Question:* The distance between New York and Paris is 3,625 miles. If we could travel the speed of light between New York and Paris, how many one way trips could we make in 1 second?

Answer: roughly $\dfrac{186000}{3625} = 51+$ trips between the two cities in one second!

The speed of light is an actual speed that light travels.

*Question:* If we make only one trip between the two cities, how long would it take us, if we travel the speed of light?

Answer: $\dfrac{3625}{186000} = 0.019489$ seconds,

or roughly 2 hundredths of a second. If we could travel the speed of light for $0.019489$ seconds we could travel the equivalent distance from New York to Paris.

*Question:* Can you write this equation using the symbol $c$ for speed of light? (Hint: Use the above formula.)

Answer: $speed \times time = distance$ or $c \times t = distance$ or $d = ct$

This illustrates a fundamental fact; $c$, (the symbol for the speed of light), an extremely fast speed multiplied by an extremely short length of time can equal the same distance as a slower speed multiplied by a longer length of time.

*Question:* If you live on one edge of Manhattan and if you could travel the speed of light to get to a grocery store on the opposing edge of Manhattan, how long would it take you to get there if Manhattan is 13.4 miles long?

Answer: $\dfrac{13.4}{186000} = 0.000072043$ seconds or about 7 hundred-thousandths of a

second, in scientific notation $7.2043 \times 10^{-5}$ seconds. This demonstrates that the interval of time can be extremely small.

*Question:* If you could step the distance of one meter, (approximately 39 inches) at the speed of light, how long would it take you to traverse this distance?

Answer: $\dfrac{1}{2.9979 \times 10^{8}} = 3.3356 \times 10^{-9}$ seconds or roughly 3 billionths of a second (0.0000000033356 s) or a little over 3 nanoseconds.

The above equation, $speed \times time = distance$ is true given any speed ($c$), any time-length ($t$), and any distance ($d$), i.e.:

$2.9979 \times 10^{8}$ m/s $\times\ 3.3356 \times 10^{-9}$ s $= 1$ meter or in shorthand notation, consider these three related equations:

$$d = ct \qquad c = \frac{d}{t} \qquad t = \frac{d}{c}$$

Any distance can be defined by the speed of light multiplied by a time-length. The combination of symbols $ct$ is a mathematical expression for a distance.

*The definition of the physical element of **distance**, and the general equation for distance, $d$:*

$$d = ct$$

For a specific length or distance, $d = ct_d$ where $t_d$ is a specific time-length multiplied by the speed of light to yield a specific distance. The specific value determined by $ct_d$ is a subset of the physical element of distance.

Following the above example, calculating the time for light to travel one meter:

$$ct = 1m$$

$$t = \frac{1}{c}\ (s)$$

Since $c = 2.9979 \times 10^{8}$ meters per second $t = \dfrac{1}{2.9979 \times 10^{8}} = 3.3356 \times 10^{-9}$ s

$$d = ct_d = 1m \text{ when } t_d = 3.3356 \times 10^{-9}s$$

The one meter time-length, denoted $t_{1m}$, is approximately 3.3356 nanoseconds.

The $ct$ expression for the physical element of distance is "$ct$." The $ct$ expression for time is "$t$." The $ct$ expression for the physical element of speed is "$c$". Note that individual speeds can be calculated, and any speed, symbolized "$s$," can be represented as some fraction of the speed of light.

*Question:* We are members of the crew on the Starship Enterprise. Captain Kirk orders Chekov to proceed to warp 2. How is warp 2 described mathematically?
Answer: $2c$ meaning two times the speed of light.

*The definition of the physical element of **motion/speed**, and the general equation for speed, s:*

$$s = nc$$

where $n$ is a number, a fraction or ratio of the speed of light; i.e. a ratio of physical elements with the same number of factors of $c$ and $t$. When time-lengths are part of an equation this $n$ is incorporated into a time-length. For simplicity, this $n$ will usually be dropped in this presentation. The symbols $c$ and $t$ are carriers of the units; in the case of SI, meters and seconds.

*Question:* How would you calculate a speed of one meter per second (m/s) using the symbol $c$?
Answer:  $s = 1\text{m/s} = nc$

$$n = \frac{1}{c} = 3.3356 \times 10^{-9}$$

In this case $3.3356 \times 10^{-9}$ is a unit-less number, since units of m/s divided by units of m/s is 1 or unit-less.

Imagine that you are in bed lying on your back. Hold your hand up toward the ceiling. Spread your index finger and your thumb apart making a distance between your thumb and forefinger. Draw that distance in your mind. Imagine that both your thumb-tip and your fingertip can draw ink lines on the ceiling. Move your extended hand across your field of vision, drawing the imaginary lines across the ceiling. This motion creates a distance that is moving, a "moving distance."

*Question:* What two factors determine this moving distance?
Answer: Speed and distance    or    $c$  and    $ct$

*Question:* Can you write an equation for moving distance using the symbols $c$ for the speed of light and $t$ for time?

Answer: $moving\ distance = c * ct = c^2t$

The symbol $c$ can be interpreted as representing the speed of light, which it does, but it can also be verbally interpreted and read as "moving." A moving distance is a factor of speed ($c$) multiplied by a distance ($ct$), the distance being any length, large or small. For example, $c^2t = c * ct$ is the speed of light squared multiplied by a time factor, but this is also a moving distance. They are one and the same.

A reasonable name, modi, an acronym for **mo**ving **di**stance, can be assigned to this physical element.

*The definition of the physical element **modi** and the general equation for modi:*

$$modi = c * ct = c^2t$$

Imagine that we are at a stoplight in a yellow Ferrari. The light turns green, you press your foot on the accelerator pedal and off we go in a blaze of glory. Our backs get pushed into our seats. In short we experience an accelerating thrill ride. This acceleration is one dimensional; that is, it is experienced only in one direction.

*Question:* If we experience a continuous acceleration of 20 m/s², at the end of 5 seconds, how fast are we traveling?

Answer: 20 m/s² X 5s = 100 m/s

*Question:* What is the formula for this relationship?

Answer: $acceleration \times time = speed$   or    $s = at$

*Question:* Can you write a family of $ct$ equations which describes this relationship?

Answer:   $a = \dfrac{nc}{t} = \dfrac{c}{t}$     $nc = \dfrac{c}{t_1} * t_2$     $t' = \dfrac{nc}{\frac{c}{t_1}} = nt_2$

The $n$ can be incorporated into the $t$ to yield a $t'$, and all the $t$'s can be combined. This yields the simple family of equations:

$$a = \dfrac{c}{t} \qquad c = at \qquad t = \dfrac{c}{a}$$

A one dimensional acceleration, or accelerated ratio, is determined by the ratio:

$$acceleration = \frac{speed}{time}$$

In SPT, it is important to distinguish this one dimensional acceleration. This one dimensional acceleration, **accelerated _rat_io**, acronym acrat is a physical element.

*Question:* Can you guess the $ct$ expression for the physical element of acrat?

Answer: $\dfrac{c}{t}$    or    $ct^{-1}$

Acrat can be calculated and given a value:

$$n * \frac{c}{t} = acrat = \frac{c}{n't} = \frac{c}{t'} \text{ or just: } \frac{c}{t}$$

The $n$ is a ratio or number. In SPT this ratio can be a ratio of time-lengths, a ratio of distances, a ratio of areas or any ratio of two of the same physical element. The $n$ for ratio or number is incorporated into the final $t$. An accelerated ratio is an accelerating distance. However, the tense of the verb is important. An accelerated ratio refers to something that has happened. An accelerating distance, references a distance which has yet to be determined. In general, the approach employed in SPT is to refer to the physical element by the subject which has happened. This will become more clear as the theory is developed.

*The definition of the physical element **acrat** and the general equation for acrat, $\dfrac{c}{t}$ :*

$$acrat = \frac{c}{t}$$

If accelerating distance is assigned the symbol $d_{a'}$, then the general equation for accelerating distance is:

$$d_{a'} = \frac{c}{t}$$

The prime notation indicates an ultimate distance determined by time, a distance primed to happen. The physical element is named acrat which has a subset of accelerating distance. The Systeme International units of acrat are $m/s^2$ or $ms^{-2}$. An advantage of SPT is that any system of units can be used as long the time unit in speed of light is expressed in the same unit as the time unit. Example: the time unit could be "Martian day" and the speed unit could be expressed in terms of "Jupiter



*Eric Mitchell Horn*

diameter per Martian day."

An accelerating distance multiplied by a time-length yields the physical element speed.

$$\frac{c}{t_1} * t_2 = nc \quad \text{where } n = \frac{t_2}{t_1}$$

An accelerating distance multiplied by time squared yields the physical element of distance; $t_1$ and $t_2$ represent different specific time-lengths:

$$\frac{c}{t_1} * t_2^2 = ct \quad \text{where } t = \frac{t_2^2}{t_1}$$

*Question:* If you experience a one dimensional acceleration of 20 m/s² in the yellow Ferrari, how far will you travel in 5 seconds?
Answer: 20m/s² X (5s)² = 500 meters

*Question:* What is a general formula for this relationship?
Answer: $\quad acceleration \times time^2 = distance \quad$ or $\quad d = at^2$

*Question:* Can you write a $ct$ equation which describes this relationship?
Answer: $\quad ct = \frac{c}{t} * t^2 \quad$ (Note $t' = \frac{t_1^2}{t_2}$, i.e. the $t$'s can be combined into one value $t$.)

*Question:* Can you guess the $ct$ equation for a physical element determined by accelerating a distance?
Answer: $accelerated\,distance = acrat * distance = \frac{c}{t} * ct = nc^2$ where

$$n = \frac{t_1}{t_2}$$

An acronym for **ac**celerated **di**stance (acdl) Is reasonable name for this physical element.

*The definition of the physical element of **acdi** and the general equation for acdi:*

$$acdi = nc^2$$

Einstein viewed the physical element of $c^2$ as an accelerating area, which it is. But should this physical element be named for something that has not happened yet: i.e.

area? The physical elements of time and speed/motion are essential to SPT. The tense of the word accelerated refers to something that has happened, which is the approach taken in this discussion.

If $A$ is the symbol for area and $a'$ is a symbol for "accelerating," then the equation for a subset of acdi, accelerating area, is:

$$A_{a'} = nc^2$$

The ultimate area described is primed to happen if the time element is fixed. An accelerated distance and an accelerating area are one and the same.


*Question:* Can you guess what acdi multiplied by time is?
Answer: $acdi * time = modi$

$$nc^2 * t = c^2 t' = c^2 t$$

*Question:* Can you guess what acdi multiplied by two factors of time is?
Answer: $acdi * time^2 = area$

$$A = nc^2 * t^2 = nc^2 t^2 = c^2 t^2$$

Again, note that $t$ represents a general time element, with the $n$ being incorporated into the $t^2$ element.

Imagine that you are still lying on your back, with a distance defined by the space between your thumb and index finger. Move your hand drawing the imaginary lines across the ceiling. Consider the time it takes to move your hand from one side of your body to the other. The area between the imaginary lines on your ceiling depends upon when you stop the motion. If you increase the distance between your thumb and fingertip, a larger area is defined by the motion of your hand. If you increase the speed at which you move your hand, a larger area is defined for the same amount of time.  If you increase the time of motion keeping the same speed and the same distance between your fingers, the final area drawn on your ceiling will be larger.

*Question:* What three factors determine the imaginary area traced across your ceiling?
Answer: distance, speed, and time

*Question:* Can you write an equation for area using these three factors and the symbols for speed of light and time?
Answer: $A = ct * c * t = c^2 t^2$

*Question:* What is modi multiplied by time?

Answer: $modi * time = area \quad$ or $\quad A = c^2 t * t = c^2 t^2$

*Question:* What is an area divided by a time element?

Answer: $\dfrac{c^2 t^2}{t} = c^2 t = modi$

*Question:* What is an area divided by a time element squared or two independent time elements?

Answer: $\dfrac{c^2 t^2}{t^2} = \dfrac{c^2 t^2}{t * t} = nc^2 = acdi$

Hopefully, if I have presented the above concepts adequately, you are beginning to understand that all factors of time and all factors of the speed of light are separable, and can be combined in any order to result in a particular physical element, i.e.:

$$Area = A = c^2 t^2 = c * t * t * c = t * c * c * t = c * c * t * t = c * t * c * t$$

*The definition of the physical element of **area**, and the general equation for area, $A$ :*

$$A = c^2 t^2$$

If $A$ is a symbol representing an area and $l$ symbolizes length and $w$ symbolizes width, then $A = lw$. Since $ct$ can equal any length, then $A = ct_1 \times ct_2$ where $t_1$ and $t_2$ are different time-lengths.

Any area can be represented by the equation $A = (ct)^2 = c^2 t^2$ when $t = t' = \sqrt{t_1 * t_2}$

The $t'$ notation will often be omitted in this presentation and it should be understood that usually $t = t'$.

*The definition of an area of one meter squared:*

$A = 1$ m² when $t = 3.3356 \times 10^{-9}$s or $t = \sqrt{t_1 * t_2} = 3.3356 \times 10^{-9}$s

The time-length of 1 m², $t_{d_{1m}} = 3.3356 \times 10^{-9}$s

An area of $1$ m² $= c^2 t_{d_{1m}}^2$

The notation $t_{d_{1m}}$ can be read backwards starting with the subscript, "one meter distance time."

*Question:* How would you write an equation for determining a specific area given a specific distance-time?

Answer:     $A = c^2 t_d^2$

The notation, $t_d$ , represents a distance-time. The physical element is time. The subset of distance-time, $t_d$ , is used to determine a distance equal to $c t_d$ , where $t_d$ may equal the square root of two time elements multiplied together, i.e. $t_d = \sqrt{t_1 * t_2}$; e.g.:

An area of 36 m² equals:

$$c^2 t_{6m}^2 = c t_{1m} * c t_{36m} = c t_{2m} * c t_{18m} = c t_{9m} * c t_{4m} = c t_{6m} * c t_{6m}$$ and all combinations in between.

Return to you hand. Close the distance between your thumb and fingertip. A rough circle is formed, a hole circumscribed by your thumb and bent fingers. Consider the area of this circle or hole. Extend your arm and move the circle through the space above your head.

*Question:* What does this motion create?
Answer:  A moving area

*Question:* What are two factors which determine the quantity of this moving area?
Answer:  speed and area     or  $c$  and    $c^2 t^2$

*Question:* Can you write a $ct$ equation for this moving area?
Answer:   $moving\ area = c * c^2 t^2 = c^3 t^2$

A reasonable name for this physical element is moar, an acronym for **mo**ving **ar**ea.

*The definition of the physical element **moar**, and the general equation for moar:*

$$moar = c^3 t^2$$

A moving area has three factors of speed and two factors of time. The two factors of time can be independent or have different values from one another. In other words, moving area, like any $ct$ expression or $ct$ equation can be factored and recombined in any order, i.e.:

$$moar = c * t * c * t * c = c^2 t * ct = c^3 * t^2 \text{ etc.}$$



*Eric Mitchell Horn*

Three factors of $c$ and two factors of $t$ in combination will always represent a moving area. Now consider if an area is accelerated.

*Question:* Can you write a $ct$ equation for an accelerated area?

Answer:    $accelerated\ area = c^2 t^2 * \dfrac{c}{t} = c^3 t$

A reasonable name for this physical element is acar, an acronym for __ac__celerated __ar__ea.

*The definition of the physical element **acar** and the general equation for acar:*

$$acar = c^3 t$$

*Question:* What is an accelerated area multiplied by a time-length?
Answer:   $acar * t = c^3 t * t = c^3 t^2 = moar$

Now consider the volume of space defined by the motion of your hand.

*Question:* What are the factors that ultimately determine the volume defined by the motion of the circumscribed area enclosed by your thumb and index finger?
Answer: moving area and time

 *Question:* Can you write this $ct$ equation?
$Volume = c * c^2 t^2 * t = c^3 t^3$     or:     $V = moar * time = c^3 t^2 * t = c^3 t^3$

*Question:* What is an area multiplied by a distance? Can you write this $ct$ equation?
Answer:    $c^2 t^2 * ct = c^3 t^3 = Volume = c * c^2 t^2 * t = c^3 t^3$

If $V$ is a symbol for volume, then any volume can be represented as $V = l \times w \times h$. Since length, width and height are all distances, any volume can be represented by a general equation.

In SPT the symbol $V$ for volume represents a "static" volume, a volume determined by $c^3 t_V^3$ where $t_V$ is a volume-time. The space may be moving, but the volume of space moved remains the same with the passage of time. The symbol $V_S$ represents a specific space volume which is often determined by a moving area multiplied by a time-length. The symbols $S$ and $V_S$ refer to space in general or a specific volume of space. $V_S$ read backwards is space volume. These symbols may be used interchangeably. In SPT, the choice of the symbol is determined by the context and the manner in which the volume, space volume, or space is derived. In general, the symbol $V$ will be associated with gravity, and $V_S$ will be associated with a scat-volume ($V_{\varsigma}$). This will be clarified in the following chapters.

*The definition of the physical element of **space** and the general equation for space, $S$:*

$$S = c^3 t^3$$

$$V = V_S = S = (ct)^3 = c^3 t^3$$

*The definition of 1 m³:*

$$1 \text{ m}^3 = c^3 t^3_{d_{1m}}$$

If $V = 1\text{m}^3$, then $c^3 t^3 = 1\text{m}^3$ when $t = (\dfrac{1}{c^3})^{\frac{1}{3}} = 3.3356 \times 10^{-9}\text{s}$ or $t = (t_1 * t_2 * t_3)^{\frac{1}{3}}$

The time-length of 1m³, $t_V = 3.3356 \times 10^{-9}\text{s}$

Close your fist, creating a volume, a volume of space. Move your fist across the ceiling creating a moving volume of space, or "moving space."

*Question:* Can you write the $ct$ equation for "moving space?"

Answer: $moving\ space = c * c^3 t^3 = c^4 t^3$

A reasonable name for this physical element is "mosp," an acronym for **mo**ving **sp**ace.

*The definition of the physical element **mosp**, and the general equation for mosp:*

$$mosp = c * c^3 t^3 = c^4 t^3$$

Moving space involves four factors of the speed of light and three factors of time. The three time factors can all be different, and those individual time elements can be ratios of powers of time; e.g. $t_{g'} = \dfrac{t_{\varsigma}^4}{t_V^3}$ or $t_{R_K} = \dfrac{t_{\varsigma Q e}^3}{t_{\varsigma P}^2}$ or $t' = \sqrt{t_1 * t_2}$ etc.

Don't worry about the notation for now. Just be aware that many factors can be hidden within a specific time value.



*Eric Mitchell Horn*

*Question:* What if a volume of space is accelerated? Can you write an equation for accelerated space?

Answer:  $accelerated\ space = c^3 t^3 * \dfrac{c}{t} = c^4 t^2$

Acsp, an acronym for **ac**celerated **sp**ace, is a reasonable name for this physical element.

*The definition of the physical element of **acsp**, and the general equation for acsp:*

$$acsp = c^4 t^2$$

*Question:* What is accelerated space multiplied by a time-length?
Answer:  $acsp * t = c^4 t^2 * t = c^4 t^3 = mosp$

*Question:* What is accelerated space multiplied by two factors of time?
Answer:   $acsp * t_1 * t_2 = c^4 t^2 * t^2 = c^4 t^3 * t = mosp * t = c^4 t^4 = ?$
(Answer forthcoming...)

Tidying up mathematical loose ends:

$$s = nc$$

$$n'c^2 = n_1 c * n_2 c$$

$$n'c^n = (n_1 c * n_2 c * \ldots n_n c)$$

For simplicity, this $n$ will be dropped in most of this presentation, and it should be understood that for many applications $n = n'$ and this $n'$ is incorporated into a time-length when a time factor is present.

Definition of $t$ prime, $t'$:

$$t^2 = t_1 \times t_2$$

$$t' = \sqrt{t_1 * t_2} = (t_1 * t_2 * t_3 \ldots t_n)^{\frac{1}{n}}$$

Likewise, $n'$ prime; distance prime, $d'$;  area prime,  $A'$; volume prime, $V'$; etc.:

$$n' = \sqrt{n_1 * n_2} = (n_1 * n_2 * \ldots n_n)^{\frac{1}{n}}$$

$$d' = \sqrt{d_1^2 * d_2^2} = (d_1 * d_2 \ldots d_n)^{\frac{1}{n}}$$

$$A' = \sqrt{A_1 * A_2} = (A_1 * A_2 \ldots A_n)^{\frac{1}{n}}$$

$$V' = \sqrt{V_1 * V_2} = (V_1 * V_2 \ldots V_n)^{\frac{1}{n}}$$

The following basic structure, multiplying by $c$ when moving up and to the left and multiplying by $t$ when moving up and to the right, forms a beginning framework of SPT. Each $ct$ expression at the intersection of the $c$ and $t$ lines mathematically represents an equation for a physical element. These physical elements are broad categorizations and may include many subsets. It will be shown that some physical elements already have names that include all subsets; (e.g.; distance/length, area, volume/space, energy, etc.)  For other physical elements only the subsets are named and many physical elements do not have any identifiable name that includes all subsets.  Some physical elements may not have any associated names at all. Each physical element should be named in order to be precise and know exactly the subject of discussion.

For the nonprofessional audience, if you have understood the above introduction, you can understand this theory. These simple expressions form the foundation of SPT. A structure or framework can be created to depict the relationships of these physical elements. See the following page.



*Eric Mitchell Horn*

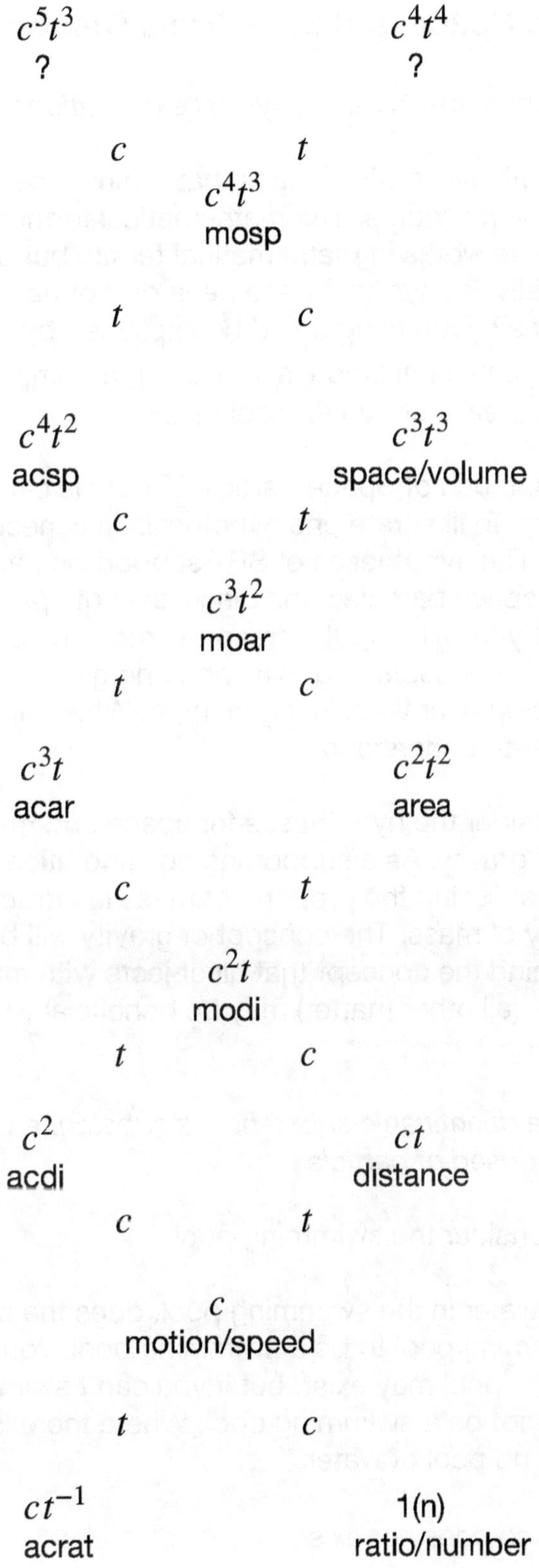

$$c^5t^3$$
?

$$c^4t^4$$
?

$c$     $t$

$$c^4t^3$$
mosp

$t$     $c$

$$c^4t^2$$
acsp

$$c^3t^3$$
space/volume

$c$     $t$

$$c^3t^2$$
moar

$t$     $c$

$$c^3t$$
acar

$$c^2t^2$$
area

$c$     $t$

$$c^2t$$
modi

$t$     $c$

$$c^2$$
acdi

$$ct$$
distance

$c$     $t$

$$c$$
motion/speed

$t$     $c$

$$ct^{-1}$$
acrat

1(n)
ratio/number

# The Space and Scat/Mass Relationship

**Hypothesis:** *Physical phenomena have physical explanations.*

This simple driving hypothesis in SPT means that ordinary people like you and I can understand how the universe works. The mathematical formulas presented by SPT describe how the universe works in mathematical terms, but physical phenomena can also be described verbally. If physical phenomena do not have physical explanations, why do we study physics? Everything could be explained by gods and the magic they possess, or by a video game simulation, a $1$ or a $0$ in a computer. Mystery solved! Time to go home! Otherwise, please continue reading.

Beyond this hypothesis, much of Space Particle Theory is circular, and a particular starting point is arbitrary. To illustrate one hypothesis, it is necessary to consider a supporting hypothesis. The hypotheses of SPT support one another. Since SPT receives its name from space particles, the discussion of space is presented first. However, it may be that you will not fully grasp the meaning of this presentation of space and mass until the concepts of gravity and energy are presented. Your understanding will develop over time. Hang in there. When all relationships are integrated, SPT is easier to understand.

It may be helpful to consider the hypotheses for space and mass which are presented below, in the context of gravity. As a supporting consideration, as far as is currently known, every object possessing the property of mass is attracted to every other object possessing the property of mass. The concept of gravity will be developed throughout this book. Keeping in mind the concept that all objects with mass (all matter) attract all other objects with mass (all other matter) may be beneficial when considering the presentation below.

**Hypothesis:** *Space is a quantifiable substance, a substance which can be divided into parts, a substance composed of particles.*

*Conceptual analogy:* Consider the swimming pool.

*Question:* If there is no water in the swimming pool, does the swimming pool exist? Answer: No! For a swimming pool to be a swimming pool, you must have water. The structure surrounding the pool may exist, but if you can't swim because there's no water in the pool, it cannot be a swimming pool. Where there is no water molecule (water particle), there is no pool of water.

**Hypothesis:** *Empty space does not exist.*

Where there is no space particle, there is no space, at least in our realm of the observable universe.



*Eric Mitchell Horn*

**Hypothesis:** *The space particle (acronym spart) is the smallest quantifiable substance in the universe.*

**Hypothesis:** *The space particle has a three dimensional size and shape.*

**Hypothesis:** *The space particle fills space; i.e. the space particle is space filling, and each space particle is in physical contact with other space particles over the entire surface area of each space particle.*

This rules out a sphere as the shape of the space particle, since a collection of spheres will always have inter-spherical spaces, regardless of how small each individual sphere is.

**Hypothesis:** *The space particle is the only true particle.*

**Hypothesis:** *Individual space particles are presently undetectable, and are likely to remain undetectable.*

Recall the definition of space:

$$S = c^3 t^3$$

If a time-length, $t$, is sufficiently small, $c^3 t^3$ is an equation for the size or volume of a space particle, symbolized $S_0$.

*The mathematical definition of the **space particle** and the equation for the size of the space particle, $S_0$ :*

$S_0 = c^3 t_{S_0}^3$     where $t_{S_0}$ is presently unknown. This volume is hypothesized in a later chapter.

This equation for the size or volume of the space particle is a general equation. It does not specify the shape of the space particle. Any shape will have an equivalent volume to that described by this equation. In SI units, $m^3$ is a general unit for volume given any shape. The space particle is a subset of the physical element of space.

*Conceptual analogy:* Consider that you are standing on the precipice in the middle of an infinitely tall waterfall. Or, consider that you are in a swimming pool, and the plug is pulled.

*Question:* Do you know where the water goes once it falls off the cliff or once it goes down the drain?

Answer: No. You only know that the water goes out of your observable realm.

**Hypothesis:** *Space particles are continually flowing out of our observable realm.*

*Supporting observation*: A black hole was first visualized telescopically in 2019. Black holes were predicted by Einstein's general theory of relativity in the early 20th century. Black holes, according to general relativity, "swallow" [consume] something called space-time. In SPT, space particles are consumed through time; i.e. consumed with the passage of time. Space particles pass out of our realm of observation.

*Question:* Where does space go once it passes through a black hole?
Answer: Unknown. It goes out of our observable realm to an unobservable realm.

*Definition: The **outer-verse** is the part of the universe which is observable or may become observable with improvements in technology.*

*Definition: The **inner-verse** is the part of the universe which is unobservable and likely to remain unobservable even with improvements in technology.*

The inner-verse can be considered to be a "black box," an unknown, undetermined entity.

**Hypothesis:** *The universe consists of an observable realm, i.e. an observable universe, henceforth referred to as the outer-verse, designated $U_o$ ; and an at present unobservable realm, an unobservable part of the universe, henceforth referred to as the inner-verse, designated $U_i$ .*

In mathematical terms, the total universe, $U_t$ , is equal to the sum of the outer-verse plus the inner-verse:

$$U_t = U_o + U_i$$

*Question:* Returning to the waterfall analogy, ignoring its height, what can determine a waterfall's size?
Answer: The volume of water that flows over the cliff.

*Conceptual Analogy:* Imagine the river, flowing toward the precipice or cliff. Consider the volume of water which forms the waterfall the instant before going over the cliff. Imagine that you have a magic knife and you can cut this instantaneous slice of water out of the river. This slice of water is extremely thin, but still has a volume.

*Conceptual analogy:* Imagine a large swimming pool. The swimming pool has a drain with a plug in the bottom of the pool. Pull the plug. The water flows out the swimming pool through the drain, and subsequently that water is lost to our observation. Take your magic knife and cut out the instantaneous slice of water the instant before it goes down the drain out of sight.

*Question:* Considering the waterfall and swimming pool analogies, is the slice of water composed of particles comparable to a slice of space?
Answer: Yes, water molecules are particles of water. Similarly, SPT considers space to be composed of space particles.

In SPT, an analogous volume to these instantaneous volumes of water is named the consumptive volume, symbolized $V_c$ , and can be described mathematically.

*The definition of the **consumptive volume** and the equation for the consumptive volume, $V_c$ :*

$$V_c = c^3 t_V^3$$

In SPT, the consumptive volume is a subset of the physical element of space, and a volume-time, symbolized $t_V$ , is a subset of the physical element of time.

*The definition of the subset of **volume-time**, and the equation for volume-time, $t_V$:*

$$t_V = (\frac{V_c}{c^3})^{\frac{1}{3}}$$

*Question:* Is the water in the above analogies moving?
Answer: Yes, the water molecules are flowing at a speed just before going over the cliff in the waterfall analogy, and they are also flowing at a speed just before going down the drain in the swimming pool analogy. Similarly, space particles are flowing at a specific speed in SPT just before being lost to our observation.

*Question:* Can you quantify the waterfall by multiplying the magic slice of water by the speed that that water is moving?
Answer: Yes.

$$waterfall = instantaneous\ volume \times speed = V_{H_2O} * s$$

*Conceptual analogy:* Consider a high volume "dish room." This is a workplace responsible for cleaning dishes that a large number of people generate at a large restaurant or university dining hall. A typical dish room will have a large food disposal unit to process the high volume of food waste. This disposal unit has a limited processing capacity. It can only process so much food per second. Imagine a room filled with food waste and you can magically move this garbage disposal while it is consuming the food waste.

*Question:* What is the maximum speed that the food disposal unit could be moved in any direction?
Answer:  You could achieve  a maximum speed which would match the maximum speed at which the garbage disposal unit processes the food waste, i.e. $V_{food} * s$, where $V_{food}$ is the "instantaneous" volume of food processed by the garbage disposal.

*Observation:* Matter cannot be accelerated to speeds beyond the speed of light.

*Question:* In SPT, what is the speed that space particles are traveling before being lost to our observation?
Answer: As in the above conceptual analogy, the maximum speed that the matter can be moved correlates with the maximum speed that the food can be consumed by the food disposal unit.  Since matter can not be accelerated beyond the speed of light, this must be the speed that space particles are traveling before being lost to our observable realm.

**Hypothesis:** *The speed of light is the speed that space particles are traveling when passing from the outer-verse to the inner-verse.*

*Question:* Considering our waterfall equation, $V_{H_2O} * s$, what is an analogous equation in SPT?
Answer:  $V_c * c$

*Conceptual analogy:* A stable waterfall does not move relative to the land mass around the waterfall. The waterfall does not move. The water that forms the waterfall moves.

*Question:* Considering the previous introductory mathematical presentation, what does $V_c * c$ represent?
Answer: $moving space = volume \times speed = V \times c = mosp$

In SPT, this equation represents a subset of moving space, a $moving volume$ designated by the acronym "mov" for **mo**ving **v**olume.

*The definition of the subset of **mov** and the equation for mov:*

$$mov = c^4 t_V^3$$
$$mov = c * V_c = c^4 t_V^3$$

                              *Eric Mitchell Horn*

Mov does not have a time element other than to define a "static" volume. A volume of space is moving, but the volume which is moved may remain in the same place in space, and the volume moved remains the same with the passage of time. The space particles which are continuously falling over the proverbial cliff are different, just as in the waterfall analogy, but the volume of the space that is moved remains the same. It will be shown that mov is a moving volume of space involved with gravity. This reflects the fact that an object possessing the property of mass may remain in the same place in space relative to other objects possessing the property of mass.

Recall the definition of the physical element moving space, or mosp:

$$mosp = c^4 t^3$$

This physical element has two subsets, an instantaneous moving volume, $mov$ as in $V_c$ , and **mo**ving **s**pace with **t**ime, acronym most, to be defined shortly. Most will have a time element associated with it. Confusion is expected at this point in the theory development. Hopefully, the relationships between mosp, mov, and most will become clear as the theory is presented. Mov and most are the two subsets of the physical element of moving space, mosp.

Returning to the waterfall analogy:

*Question:* While the above waterfall equation is a quantifying equation, does it describe how long the waterfall exists? Do we know if the waterfall lasts for 20 seconds or a million years?
Answer:  No. To describe the waterfall thoroughly we must add a time element to our waterfall equation:

$$waterfall = V_{H_2O} * s * t$$

The *t* element describes the length of time that the waterfall exists.

Likewise, for an analogous physical element in SPT, we must also add a time element.

$$? = V_c * c * t = c^3 t_V^3 * c * t$$

*Question:* What physical element does this equation describe? Can you guess?
Answer: Keep reading.

*Conceptual Analogy:* Consider that you are lying on a platform suspended just above the waterfall, able to physically touch the slice of water, just before it falls through the air. If the flow of the water is sufficiently slow, you can easily put your hand through the water. In your mind, speed up the river, making it flow very fast.

*Question:* Can you put your hand through the water as easily as before?
Answer: No.

*Conceptual analogy:* Imagine the speed of the water coming out the nozzle of a water cannon.

*Question:* Can you put your hand through the flow of water emanating from a nozzle of a water cannon?
Answer: It is very difficult.

*Question:* Can an extremely fast flow of water molecules serve as an analogy for extremely fast flowing space particles?
Answer: Yes.

*Question:* Can an extremely fast flow of water molecules serve as a conceptual analogy for matter?
Answer: Yes! Suppose the speed of the flowing space particles is near the speed of light. Can you put your hand through a steel door?

*Question:* What is the question mark, "?", in the $V_c * c * t$ equation?
Answer:   Matter !!!

**Hypothesis:** $matter = V_c * c * t$

**Hypothesis:** $V_c * c * t$ is an equation that quantifies matter, it describes what matter **is.**

**Hypothesis:** Matter, an object with the property of mass, is touchable moving space through time. Note: $c * V_c * t$ can be read as "moving space through time," i.e. :

$$c = moving \qquad V_c = space \qquad t = time$$

*Question:* What is the derived SI unit for a quantity of matter in SPT?
Answer:   $V_c = m^3$    $c = m/s$    $t = s$ ;    $m^3$ x $m/s$ x $s = m^4$
The derived SI unit for a quantity of matter is the $m^4$. The SI unit of mass (rest-mass) is the kilogram. A kilogram and a meter$^4$ will be shown to be related by a ratio of time-lengths in an upcoming chapter.

**Hypothesis:** The difference between the space we can touch; i.e. ordinary matter, and the space previously referred to as "empty space" is the velocity of the space particle flow at the touch point.

**Hypothesis:** *Matter is formed from the flow of space particles which are traveling near the speed of light as the space particles approach an edge to the outer-verse.*

*Question:* If space particles pass from the outer-verse to an inner-verse, what is the structure through which the space particles pass?
Answer: Unknown. At this stage of development of SPT, we can only say that these structures are a complex of holes which connect our outer-verse to an inner-verse. This is another "black box" in SPT theory.

*Definition:* A **hole-complex** *is a complex of holes through which space particles pass between the inner-verse and the outer-verse.*

**Hypothesis:** *A hole-complex does not need to move through space for a moving volume of space to exist.*

For SPT to work as a model for the universe, two conditions must be met. The pool water (space), must be drained someplace; that place being unobservable, and the pool water (space) must be replaced. In SPT these outlets and inlets are holes or hole-complexes through which space particles flow or pass between the outer-verse and the inner-verse. The hole-complexes are embedded in, or are part of a boundary between, the outer-verse and the inner-verse.

**Hypothesis:** *The outer-verse, $U_o$, and the inner-verse, $U_i$, are connected physically to one another via hole-complexes.*

**Hypothesis:** *Space particles flow through a network of hole-complexes which connect the inner-verse to the outer-verse.*

SPT does not rule out the possibility of multiple unobserved inner-verses. In fact, if space is conserved, a second inner-verse may be necessary, a subject considered in the discussion section of this book.

**Hypothesis:** *The space particle is the currency of exchange in the universe.*

**Hypothesis:** *The space particle has no mass.*

A space particle, in and of itself, is not matter. Matter is formed from a volume of space moving extremely fast through time. If space is not moving, it is not matter, and therefore cannot possess the property of mass. Matter possesses the property of mass. Einstein's relativity mass considers the property of mass.

Current physics theory uses the property of mass to quantify matter. The SI unit currently used to quantify matter is the kilogram (kg). Mass, a property of matter, can be defined as a resistance to change in motion of matter, and referencing mass as being a volume of space traveling at the speed of light is inappropriate. By Einstein's relativity formula, which considers the property of mass, the mass of an object of matter changes relative to its speed, or surrounding gravitational field, where $m_{rel} = \dfrac{m_0}{\sqrt{1 - \dfrac{v^2}{c^2}}}$. Since mass changes with changing conditions, mass is a poor quantifier of matter. Because the word matter has traditionally referenced an object that can be touched, or something which possesses the property of mass, the use of the word matter would also appear to be inappropriate when referencing this physical element. An alternative word, scat, a symbolized name for space multiplied by the speed of light and time, $Sc$ **a**nd $t$, will be used in this text.

**Hypothesis:** *Rest mass, a property that quantifies matter, can be represented by the following equation:*

$$restmass = m_{m^4} = V_c * c * t = scat.$$

The notation "$m_{m^4}$" can be read as "mass expressed in meters to the forth," or "mass in m4."

*The definition of the physical element of **scat** and the general equation for scat, symbolized $\varsigma$ :*

$$scat = \varsigma = c^4 t^4$$

$$scat = \varsigma = c^4 t_V^3 * t_{g'} = c^4 t_\varsigma^4$$

$t_{g'}$ is a specific time-length associated with the gravitational constant and will be derived later.

$$restmass = scat = c^4 t^4 \qquad where \qquad t^4 = t_V^3 * t_{g'}$$

Scat can be considered to be a volume of space moved a distance:

$$\varsigma = c^3 t^3 * ct = c^4 t^4$$

In SPT, the $t$ of $t^4$ is defined to be a scat-time, symbolized $t_\varsigma$ .

*The definition of **scat-time**, and the equation for scat-time, $t_\varsigma$ :*

$$t_\varsigma = \left(\frac{c^3 t_V^3 * c * t_{g'}}{c^4}\right)^{\frac{1}{4}} = \left(\frac{c^4 t^4}{c^4}\right)^{\frac{1}{4}}$$

A scat-time is a composite time and may be comprised of many time elements.

There are two equivalent equations for matter:

$$matter = restmass = c^3 t_V^3 * c * t_{g'} = c^4 t_\varsigma^4$$

This means that the physical element of mosp (moving space) has at least two subsets:

$$moving\,volume = mov = c^4 t_V^3$$

$$moving\,space\,with\,time = most = \frac{c^4 t_\varsigma^4}{t_\varsigma} = c^4 t_\varsigma^3$$

*The definition of the subset of **most**, and the general equation for most:*

$$most = c^4 t_\varsigma^3$$

$$most = \frac{scat}{scattime} = \frac{\varsigma}{t_\varsigma} = \frac{c^4 t_\varsigma^4}{t_\varsigma} = c^4 t_\varsigma^3$$

Both $mov$ and $most$ will have units of m⁴/s.

$$mov = V_c * c = \text{m}^3 \text{ X } \text{m/s equals } \text{m}^4\text{/s}$$

$$most = \frac{c^4 t^4}{t} = \text{m}^4\text{/s}$$

*Question:* Considering the waterfall analogy, what does this equation represent?

$$V_{H_2O} * s = \frac{waterfall}{t}$$

Answer: An instantaneous waterfall, a waterfall in which the water has not had a chance to fall over the cliff, a picture of the waterfall, not a movie of a waterfall, a waterfall without a time element....

*Question:* Likewise, what does the equation $V_c * c = \dfrac{\varsigma}{t_{g'}} = \dfrac{c^4 t_\varsigma^4}{t_{g'}}$ represent?

Answer: Rest mass without a time element, an instantaneous picture of rest mass.

*Question:* If you could take away all time, $t_{g'}$, from a rest mass, would you see anything in your picture?
Answer: I think the answer is no. Think about it.

$mov$ is analogous to the instantaneous volume of water flowing at a speed the instant before falling off the cliff.

Consider the physical element of time, $t$. It will have at least two subsets:

$$volumetime = t_V$$

$$scattime = t_\varsigma$$

The difference between the two involves the detection of motion. The volume-time involves a "static" volume, a volume of space described initially as a distance cubed; i.e. $(ct_V)^3$. A volume-time can be conceived as an instantaneous time.

*Question:* Is a volume-time a time-length without time?
Answer: Yes. It's a puzzle; isn't it? A volume-time is a strictly mathematical path to describe a specific volume, an instantaneous volume. This may be a difficult concept to comprehend when first encountered. The concept of the physical element of time having at least two distinct subsets will become easier as you continue reading.  In general, the concept of volume-time is only important in understanding gravity and its associated concepts.

A scat-time is not an instantaneous time. A scat-time is the time that we experience in our daily lives. For motion or speed to be detected, there must be a ratio multiplied by a time value. A scat-time is described initially from $scat$ which includes a time factor, a gravitational time element.  This will be described and numerically calculated in an upcoming gravitational chapter.

The physical element of space also has at least two subsets, a "static volume," $V_c$, and and a quantity of space determined from moving area through time or scat per distance:

$$c^3t^2 \times t = c^3t^3 \qquad \text{or} \qquad \frac{scat}{distance}$$

*The definition of the physical element of **space**, and the general equation for space, S:*



*Eric Mitchell Horn*

$$S = c^3 t^3$$

The definition of the subset of **scat-volume** and the equation for scat-volume, $V_\varsigma$:

$$V_\varsigma = \frac{c^4 t_\varsigma^4}{c t_\varsigma} = c^3 t_\varsigma^3$$

**Hypothesis:** *Any process in which motion is observed will involve a scat-time.*

The derived SI unit of scat is a m⁴; i.e. four physical dimensions. However, we only acknowledge that we experience three physical dimensions. This issue will be addressed later in SPT. It is important at this stage of development of SPT to resist prejudging its relevance based upon our perception of physical dimensions.

**Hypothesis:** *The property of mass results from a net volume of space, $V_c$ (consumptive volume), traveling at the speed of light through a hole complex as long as the hole-complex exists.*

**Hypothesis:** *The property of mass measures the energy required to change a particular consumptive pattern.*

Reference will often be made to an object with mass, meaning matter which possesses the physical property of mass. By hypothesis, the specific velocities of space particle flows form matter. Scat is a true quantifier of matter. In this text, the word mass and the symbol $m$ to represent mass will generally be reserved to denote the property of mass, referenced in Einstein's relativity equation. With conditions of slow speed and low gravity:

$$mass \approx scat$$

$$restmass = m_{m^4} = scat = V_c * c * t \quad (t = t_{g'}\text{; further development to follow})$$

In the remainder of this text, the small Greek letter sigma, $\varsigma$, symbolizes scat. Since the word mass and the concept of mass are in common usage, the physical element of scat will often be referenced as scat/mass in general discussion. However, it should be understood that the true entity is scat or rest mass. It will be shown that using the symbol $m$ for mass in at least one accepted equation is incorrect. The concept of scat is introduced for purposes of precision, and to convey the fact that scat is an equation for calculating rest mass. For the rest of this discussion, unless specifically noted by the notation $m_{kg}$ or kg, all rest masses are expressed in units of m⁴, specifically noted by the symbol $\varsigma$ and all equations assume a rest mass or connection to a mass expressed in $m^4$. The concept of matter being formed from a volume of space moving toward a hole-complex raises the question of space and its structure.

*Conceptual analogy:* Consider a huge block of Swiss cheese. The outer-verse, the observable realm, is the physical cheese.

*Conceptual analogy:* Consider that a chain link fence separates the physical cheese from the spaces in the cheese. The hole-complexes are part of the fence, the boundary between the physical cheese and the interior spaces in the cheese.

*Conceptual analogy:* The interior of the cheese has space, commonly referred to as "holes" in the cheese. These "holes" in the cheese are analogous to the unobservable realm or inner-verse in SPT. They are the places where space particles travel to when passing from the outer-verse to the inner-verse.

*Observation:* The volume of space appears to be getting larger at an increasing rate, a reference to observations which lead to a proposal for the existence of dark energy.

*Question:* How can the rate of the space volume expansion, increase with the passage of time, if space particles are continuously flowing to an inner-verse?
Answer: Obviously, some process is responsible for replacing and even increasing the volume of space that is being consumed by the process which creates matter. This process of replenishment of space will be considered in the next chapter.

*Question:* If space particles are being replaced continually, how is this accomplished without creating a totally chaotic outer-verse in the process?
Answer: Continue reading.

*Conceptual analogy:* Consider a glacier. The individual snow flakes, once deposited on a glacier, do not travel through the glacier. They flow with the glacier and become part of the glacier.

*Question:* Can a glacier be analogous to space?
Answer: Yes!

*Question:* Are snowflakes analogous to space particles?
Answer: In a sense, yes. A glacier is a flowing solid. It is solid ice that flows to an edge. The edge eventually breaks off, but the glacier remains, renewed by the snow that is deposited. Similarly space particles are "deposited" or "injected" into space, and become part of space, with space continually flowing to an edge.

**Hypothesis:** *Space is the equivalent of a flowing solid. The flow of space is caused by the consumption of space by an object with mass.*

If a space particle becomes part of space in an analogous manner to a snowflake becoming part of a glacier, then matter can be described as a flow of space particles to an edge, the edge being a hole through which space particles pass at the speed of light to an inner-verse. The space particles, once deposited, become part of space and flow with the space particle flow. Thus space, like the glacier, can be considered analogous to a flowing solid. Matter is formed from the extremely fast flow of space toward a hole-complex in space.

**Hypothesis:** *Individual space particles do not travel through space. Only holes or hole-complexes which connect the outer-verse to the inner-verse can travel through space.*

There are other possible analogies. Consider the "Pac-Man" game from 1980. Space particles may not be flowing out of the observable universe. Perhaps they are crushed or eaten by a subatomic Pac-Man. It is also possible that the space particles, once captured by a subatomic structure are eroded by time. But what physical process could explain this? The reverse process could be the equivalent of popcorn. space particles suddenly popping into existence, forming space foam.

Although these may be valid alternatives to an unobservable part of the universe; i.e. an inner-verse, why choose these options, especially since black holes have been observed telescopically in 2019. This observation provides visual evidence of the existence of holes in the structure of space through which a hypothesized quantity called "space-time" passes out of the observable universe. The physical idea that space particles flow through holes or hole-complexes connecting an unseen inner-verse to an observable outer-verse is simply a submicroscopic extension of a black hole.

An underlying assumption in the presentation of SPT is that a physical reality exists and that there is one and only one physical reality. There are different ways of viewing that reality, different lenses to observe that reality, but still, one and only one true physical reality. Physical reality is discernible and knowable to a point, but may not be discernible beyond a point. Many, including myself, may be troubled by the concept of an inner-verse. And yet, is not the concept of an unobservable inner-verse less troubling than a trillion-trillion-trillion little pac-men chewing up the universe, processing and compressing the space consumed in their digestive systems to nothing? The idea that time might erode the space particle may be more appealing, but what is the physical process by which this erosion of space could occur?

In summary, the quantity of mass and the property of mass are two different entities. The symbol "$m$" should not be used to represent both. The quantity, rest mass, can be expressed as a moving volume of space through time. Einstein's relativity considers the property of mass. The property of mass is a poor quantifier of matter, since the property of mass changes with changing conditions. In the next section, we will consider the quantity of rest mass; i.e. scat, and its relationship with energy.

# Energy: Part 1

From the previous space and scat/mass discussion, SPT hypothesizes:

$$restmass = scat = V_c * c * t$$

Matter is formed from a moving volume of space through time.

*Conceptual analogy:* Our swimming pool is being drained, and we need to replace the water to keep swimming. Likewise, if space in our outer-verse is being drained to an inner-verse, it must be replaced. What physical element; force, energy, temperature, area, etc. could be responsible for replacing the space consumed by a quantity of matter?

*Observation:* Space does not disappear over time. In fact, the volume of space appears to be increasing with time. This is a reference to "dark energy."

*Question:* Does energy replace the space which is consumed in the formation of matter?
Answer: Yes.

*Question:* Can dark energy be explained by SPT?
Answer: Yes!

*Conceptual analogy:* Let us consider the swimming pool. Water in the pool serves as the equivalent of space. The pool water is drained via multiple drainage outlets; the hole-complexes that are responsible for the formation of matter. For the volume of pool water to increase with time, multiple inlets need to be added to our pool model. Now suppose the swimming pool has multiple inlets for replacement of the drained water. To account for the increase in pool volume, water must be added to the pool at a faster rate than water leaves the pool.

*Question:* Is it possible to keep the swimming pool filled, even with the multiple unplugged drains? Can enough inlets be added to the pool to overcome the amount of water being drained from the pool to actually expand the volume of the pool?
Answer: Yes. It is a necessary requirement to have more inlets than outlets or have greater flow through the inlets than the outlets or some combination thereof.

*Question:* Can the pool analogy be used to explain energy?
Answer: Yes, but first we need to understand what energy actually is.

*Conceptual analogy:* Observe boiling water.

*Question:* What is the physical process that creates water vapor within the liquid water?  Is energy added to the pot?
Answer: Yes, energy is added to the pot.

*Question:* Did energy add space to the pot?
Answer: Yes. Space is continually replaced by a 3D depositional process involving energy.

*Question:* If so, what is energy according to SPT theory?
Answer: See the hypothesis below.

**Hypothesis:** *Energy is a volume of space squared per time squared.*

$$E = \frac{V_S^2}{t^2} = \frac{S^2}{t^2}$$

*Question:* How is this equation arrived at?
Answer: Suppose energy can be represented as a "flipped" hole-complex. Instead of consuming space from the outer-verse, space is returned to the outer-verse.

By previous hypothesis:

$$matter = V_c * c * t$$

$$scat = \varsigma = c^4 t_\varsigma^4$$

Applying Einstein's energy equation, $E = mc^2$ to the scat hypothesis yields:

$$E = mc^2 = c^4 t_\varsigma^4 * c^2 = c^2 \varsigma$$

Therefore, by arithmetic substitution:

$$E = V_c * c * t * c^2 = c^6 t_\varsigma^4$$

*The definition of the physical element of* **energy,** *and the general equation for energy,* *E:*

$$energy = E = c^6 t^4$$

Energy is the name given to the physical element of $c^6 t^4$. It includes all subsets; e.g. mechanical, heat, electrical, etc. Energy will always have six factors of the speed of light and four factors of time.

The derived SI units for energy are:

$$c^6 t_s^4 = \text{m}^6/\text{s}^6 \times \text{s}^4 = \text{m}^6/\text{s}^2 = (\text{m}^3/\text{s})^2$$

or energy is $(\text{m}^3/\text{s})^2 = (\dfrac{V_S}{t})^2 = \dfrac{V_S^2}{t^2}$.

Einstein's $E = mc^2$ can be easily derived by SPT. Any one dimensional acceleration can be represented by $acrat = a = \dfrac{c}{t}$. Substituting the $ct$ expression for acrat, $\dfrac{c}{t}$, in Newton's $F = ma$ yields:

$$F = ma = m * \dfrac{c}{t}$$

The well known expression for energy or work is:

$$work = energy = force \times distance$$

Any distance can be represented by $distance = d = ct$. By arithmetic substitution:

$$E = Fd = mad = m * \dfrac{c}{t} * ct = mc^2$$

The time units cancel. Combining the two above relationships yields $E = mc^2$.
$E = mc^2$ is confirmed if the time-lengths are equal, otherwise, $E = nmc^2$

*Observation:* The equation for energy implies a 6th dimension, i.e. $\text{m}^6/\text{s}^2$. This means, since our daily experience involves three physical dimensions, the process of energy results in the transfer of a volume of space per second to the outer-verse, and a volume of space per second to the inner-verse, or some alternative. This will be expanded upon in the discussion section of the book. This observation leads directly to the following four hypotheses.

**Hypothesis:** *Energy is a volume of space squared per time squared. Energy results in a volume of space per time returning to the outer-verse from the inner-verse.*

**Hypothesis:** *All massless particles that are detectable are reverse hole-complexes through which a net flow of space particles pass from the inner-verse to the outer-verse. All detectable massless particles are delivery vehicles for a volume of space returning to the outer-verse.*

**Hypothesis:** *The return of space particles is ongoing, with a specific volume of space returning to the outer-verse for every individual wavelength of electromagnetic energy.*

**Hypothesis:** *Matter and energy are equal, but opposite processes.*

The difference between matter and energy is the direction of space particle flow. When matter is "converted" to energy, the hole-complexes are disconnected from each other, the holes are "flipped," and the flow of space particles from the outer-verse to the inner-verse is reversed. Matter can be viewed as energy exiting the outer-verse to the inner-verse via Einstein's equation, $E = mc^2$; i.e. $E_{oo} = \varsigma c^2$ where $E_{oo}$ is energy transferred out of the outer-verse.

**Hypothesis:** *Matter is the physical manifestation of energy leaving the outer-verse.*

*The definition of the subset of energy, **E**nergy **o**utgoing from the **o**uter-verse, $E_{oo}$ :*

$$E_{oo} = c^2 * c^4 t_\varsigma^4 = c^6 t_\varsigma^4 = c^2 \varsigma$$

The law of conservation of matter and energy is translated as follows:

**Hypothesis:** *Hole-complexes can be rearranged, disconnected from one another and "flipped," but individual holes which connect the inner-verse and the outer-verse cannot be destroyed.*

If space is conserved, then the inner-verse is complex and must involve an inner-verse to an inner-verse or the simultaneous sharing of a common inner-verse by the outer-verse and a first inner-verse, or two identical outer-verses sharing an inner-verse. At this stage of development of SPT, it is not necessary to totally describe the structure or the total dimensions of the involved inner-verses. Other interpretations are included in the latter part of the discussion section of this book.

If space is continually replaced by an energetic 3D depositional process, the concept of space particles traveling through space in all different directions becomes chaotic and impossible. But this isn't the case if space is viewed as a solid. If you are confused, please refer to the glacier analogy in the previous chapter.

*Question:* If space is solid, how can matter move?
Answer: Matter is formed from the process of a consumption of space particles which travel through hole-complexes. The individual space particles that form matter do not move with matter. Matter is able to be moved through space only because matter results from a hole-complex in space. The hole-complex moves through space.

*Conceptual analogy:* Imagine a horse eating its way through a field of mile high grass. If the horse did not eat as it moved, it could not move. If the horse wishes to move in a specific direction, it must eat with a sustained consumptive pattern in the direction that it wishes to move.

**Hypothesis:** *Movement of matter is possible only because the space surrounding matter is consumed by the hole-complex which forms matter.*

**Hypothesis:** *Energy allows for or causes a change in a consumptive pattern.*

**Hypothesis:** *A change in a consumptive pattern allows for or causes a change in movement.*

**Hypothesis:** *A sustained directionalized consumptive pattern results in movement of matter.*

Energy, or more specifically,  the $\sqrt{E} = \dfrac{V_S}{t}$, a volume of space per time in the outer-verse, replaces the space being consumed. The direction of consumption and a corresponding replacement of space on the backside of the hole-complex allows for movement in the direction of motion.

*Observation:* The effects of a particle flow entering or encountering another particle flow will be observed upstream from the point of interaction between the particle flows.

*Conceptual Analogy:* Imagine that we are in rush hour traffic on a freeway. Cars are entering the freeway from an entrance ramp. You must decelerate to avoid an accident. A hundred cars before us have also decelerated. There's a back-up on the freeway, a negative acceleration.

*Question:* Is there an analogy between a back-up of cars on a freeway and a back-up of space particles?
Answer:  Yes. The back-up "grows" upstream from the intersection of the car particle flows.

*Question:* What is a theory equivalent of this phenomena?
Answer: Energy, $\sqrt{E}$, a volume of space per time, entering a gravitational flow stream.

*Conceptual analogy:* Consider a burning candle.

*Question:* Why does a candle flame rise? Are space particles entering a space particle flow stream?
Answer: Yes. The energy released by the burning candle is adding a volume of space ($m^3/s$) to the surrounding space particle flow stream. This causes a back-up which ultimately results in the tendency of the candle flame to rise. This phenomena could be

tested in the International Space Station. Although the gravitational field is weaker 254 miles up, the average altitude of ISS, the tendency for the candle flame to rise away from the earth, should still be observable.  This tendency will be negated, however, as a spacecraft approaches the earth-moon Lagrange point with the sun's gravity having a greater effect as the spacecraft travels away from the earth. The Lagrange point is the point at which gravitational forces of attraction between the earth and the moon are equal. As a space craft approaches this point, the candle flame should orient to oppose the sun's gravitational field. This reveals a potential experiment which could conceivably support SPT. The process of gravity is considered in the following chapter.

# Gravity: Part 1

*Question:* What is the simplest path to a physical explanation of gravity? How is it that every thing composed of matter is attracted to every other thing composed of matter?

*Conceptual Analogy:* Water is composed of particles; i.e. water molecules. Perform a mind experiment involving four fish in a water filled balloon. The fish are surrounded by a  media; i.e. water molecules. Imagine the fish consuming the water molecules, with the water molecules disappearing after being consumed by the fish.

*Question:* Do the fish appear to be attracted to one another?
Answer: Yes.

*Question:* Do the fish physically become closer to one another?
Answer: Yes.

*Question:* Could gravity be a result of the consumption of the media surrounding matter?
Answer: Yes.

*Question:* What is the media surrounding matter?
Answer: Space surrounds all matter.

*Question:*  Could gravity be the result of a consumption of space particles?
Answer: Yes. SPT is the equivalent of the old ether theory with a twist. The ether is space itself.

*Conceptual Analogy:* Imagine that you are in a large swimming pool with some friends, (hopefully close friends, because you will be very close when this mind experiment ends.) Now, put multiple drains in the swimming pool. Imagine the sides of the pool are elastic, and can contract or conform to the amount of water left in the pool. Imagine also that the drains are able to move with the sides.

*Question:* Can the drains be analogous to holes through which space particles pass?
Answer: Yes.

*Conceptual Analogy:*  Change yourself and friends into 3D drains; i.e. drains that consume the the water equally from all directions simultaneously.

*Question:* Who are you touching when the water is all drained?
Answer: It depends.

*Question:* Are you touching the friends that were physically closest to you when the plugs were pulled?

Answer: Maybe or maybe not.

*Question:* What are the factors that determine which friendly drains you are drawn to?
Answer: One factor is the relative size of your friends. If you are the size of Jupiter and all your friends are the size of Mars, then you will be the big fish in the pond. All your friends will be drawn to you. The size of the drain matters. If on the other hand, all the drains are about equal in size, the drains that you ultimately are touching depend mainly upon the distance between you and your friendly drains. You will be touching the friends that were closest to you.

**Hypothesis:** *Gravity is the observed result of a process whereby space, which is composed of space particles, travels toward hole-complexes in the structure of space.*

*Conceptual Analogy:* Consider that you are swimming across a river. A river is composed of particles of water, the net movement of these particles is the speed of the water flow of the river. Swim directly across the river, maintaining your body's perpendicular position relative to the river's flow.

*Question:* If the river is two miles wide; and your swimming speed is two miles per hour, what formula can you use to determine how long will it take you to swim across the river?

Answer: $distance = speed \times time$     or     $time = \dfrac{distance}{speed}$

$$d = 2 \text{ miles} \quad speed = 2 \text{ mph} \quad t = ?$$

$$t = 1 \text{ hour}$$

*Question:* If the speed of the river is two miles per hour; how far will the flow of the river carry you downstream?
Answer: Again: $distance = speed \times time$   $d = ?$   $speed = 2 \text{ mph}$   $t = 1 \text{ hour}$

In this case: $d = 2 \times 1 = 2$ miles.

*Question:* Is the river water's flow translatable to the process involved with gravity?
Answer: Yes.

*Question:* Will a river of water particles serve as an analogy for a river of space particles?
Answer: Yes.

*Question:* Could the flow of a river serve as an analogy to explain the arc, or orbit, of a

planet around the sun, or a satellite circling the earth?
Answer: Yes. For a theoretically perfect circular orbit of 120 miles above the earth's surface, the orbital velocity is calculated to equal 17,400 mph. To maintain that distance from the earth's surface, the speed of the spacecraft must equal a counteracting flow.

*Conceptual analogy:* Imagine that you are dropped into the middle of a river. If the river is flowing at 2 mph in one direction, and you wish to maintain your position relative to the river bank, you would have to swim at 2 mph in the opposite direction.
In the analogous case of the satellite orbiting the earth, the earth becomes the reference point on the riverbank. The maintenance of a satellite's distance from the surface of the earth is analogous to the maintenance of the swimmer's position with reference to the riverbank. Regarding the satellite's circular orbit, each point on the orbit is the same, (consider the earth to be an absolutely perfect sphere.) To maintain the distance of 120 miles above the earth's surface, and return to the exact same place relative to the earth after each orbit, the speed of the spacecraft multiplied by the time to complete one circular orbit must equal the speed of the space particle flow multiplied by the time for the spacecraft to complete one orbit. Therefore, the orbital speed must equal the speed that the space particles are traveling toward the earth at that distance from the earth.

*Question:* Does this mean that the speed of a space particle flow toward the earth at 120 miles from the surface of the earth is 17,400 mph?
Answer: Yes!

*Question:* Is gravity a force?
Answer: Yes, but with a caveat. SPT proves once and for all that gravity and force belong to the same physical element. The following is a derivation of a family of equations for determining the gravitational force between two objects with mass.  The following symbols for the given meanings will be used in this mathematical presentation.

$g$ $\quad=\quad$ gravitational force in general

$g'$ $\quad=\quad$ the gravitational force exerted by one object with mass

$g''$ $\quad=\quad$ the gravitational force acting between two objects with mass

$a_g$ $\quad=\quad$ gravitational acceleration

$k_{g'}$ $\quad=\quad$ the acceleration constant due to a single object with mass expressed in SI units of m/s$^2$

$k_{g''}$ = the gravitational constant for calculating the gravitational force of attraction between two objects with mass expressed in m⁴

$F_g$ = Newton's gravitational force expressed in newtons involving two objects with mass expressed in kilograms

$G_{kN}$ = Newton's gravitational constant, $G$ (The symbol $G$ is reserved for conductance.)

Assuming gravity is a subset in the same physical element as force, and employing $ct$ analysis, gravity must have a connection with Newton's second law of motion, $F = ma$. From the previous discussion, any accelerating distance can be calculated and symbolized by the $ct$ expression of $\dfrac{c}{t}$. If an object with mass has units of m⁴, then:

$$F = c^4 t^4 * \frac{c}{t} = c^5 t^3$$

The derived SI units of force will be:

m⁴ x m/s x 1/s = m⁵/s²

The acceleration due to gravity, $a_g$, with a constant symbolized $k_{g'}$, also has a $ct$ expression of $\dfrac{c}{t}$. Performing a $ct$ analysis on the gravitational equation and using scat/mass:

$$g_1' = \varsigma_1 * a_g$$
$$g_2' = \varsigma_2 * a_g$$
$$\varsigma_1 \varsigma_2 = \frac{g_1' g_2'}{a_g^2}$$
$$c^4 t_{\varsigma_1}^4 * c^4 t_{\varsigma_2}^4 = \frac{c^5 t_1^3 * c^5 t_2^3}{a_g^2}$$

Not specifying a specific $t$ yields the $ct$ expression:

$$c^8 t^8 = \frac{c^{10} t^6}{a_g^2}$$

For this to be true, $a_g^2$ must be $\dfrac{c^2}{t^2}$. Therefore, $a_g$ can be written as $\dfrac{c}{t}$ with the SI derived unit of m/s². The gravitational equation for one mass, symbolized $g'$, can be written as:

$$g' = \varsigma * a_g = \varsigma * \frac{c}{t_{g'}} = \varsigma * k_{g'}$$

where $t_{g'}$ is a gravitational time constant associated with the gravitational acceleration associated with one mass. The necessity of the distinction with the one prime notation will become obvious as the theory is developed.

$$a_g = k_{g'} = \frac{c}{t_{g'}}$$

Recall the definition of scat :

$$scat = \varsigma = c^4 t_\varsigma^4$$

A major usefulness of this definition lies in the fact that $c^4 t_\varsigma^4$ has a square root, i.e.:

$$\sqrt{c^4 t_\varsigma^4} = c^2 t_\varsigma^2$$

*The definition of the subset of* **scar** *(**sc**at-**ar**ea) and the equation for scar,* $A_\varsigma$ *:*

$$A_\varsigma = c^2 t_\varsigma^2$$

A scat-area is a subset of the physical element of area. The square root of a scat/mass is an area which is critical to SPT, and will allow greater understanding of the inner workings of the universe. The derived SI unit for a scat-area is the same as any area, a m². A scat-area is a moving distance multiplied by a time element:

$$scar = scat\text{-}area = A_\varsigma = c^2 t * t = c^2 t_\varsigma^2$$

Consider Einstein's $E = mc^2$:

$$E = c^6 t_\varsigma^4$$

$$E = (c A_\varsigma)^2$$

$$E = g'ct_{g'} \qquad \text{(Equivalent to } E = Fd \text{ when } g' = F \text{ and } d = ct_{g'}\text{)}$$

$$A_{\varsigma}^{2} * c = g't_{g'} \qquad \text{(Both sides translate to maximum momentum.)}$$

$$\varsigma = \frac{g't_{g'}}{c} = \frac{g'}{a_g}$$

$$g' = \varsigma\frac{c}{t_{g'}} = \varsigma a_g \qquad \text{(Equivalent to } F = ma\text{)}$$

The calculation of accelerating distance due to gravity, $a_g$:

$$a_g = k_{g'} = \frac{c}{t_{g'}} = 6.6741 \times 10^{-11} \text{ m/s}^2$$

*The definition of the **gravitational time constant** for one object with mass, and the equation for the gravitational time constant for one object with mass, $t_{g'}$ :*

$$t_{g'} = \frac{c}{a_g} = 4.4919 \times 10^{18}\text{s}$$

$a_g$ has a $ct$ expression of $ct^{-1}$ or $\dfrac{c}{t_{g'}}$. $t_{g'}$ can be calculated by a number of methods.

$t_{g'}$ is equal to the numerical value $\dfrac{1}{m_{m^41kg}}$ (seconds).

$t_{g'}$ can be calculated without considering a mass to be a m⁴:

$$P_{max} = m_{kg} * c = F_{g1kg} * t_{g'}$$

$$t_{g'} = \frac{m_{1kg} * c}{F_{g1kg}} = \frac{P_{max}}{F_{g1kg}}$$

$$P_{max1kg} = c * 1 \text{ kg}$$

$$\frac{P_{max1kg}}{t_{g'}} = F_{g1kg} = 6.6741 \times 10^{-11}\text{N}$$

$$t_{g'} = \frac{c}{6.6741 \times 10^{-11}} = 4.4919 \times 10^{18}\text{s}$$

Considering Newton's gravitational equation:

$$F_g = G_{kN}\frac{m_1 m_2}{d^2}$$

In SPT:

$$g' = \frac{c}{t_{g'}d^2} * \sqrt{\varsigma_1 \varsigma_2}$$

where $d^2$ is unit-less.

In SPT, the units are carried by the $ct$ expression, i.e.
$8\text{m} = 8 * c * t$ when $t = 3.3356 \times 10^{-9}\text{s}$

Because everything can be expressed as a $ct$ expression/equation in SPT, any numerical value can be separated from the time element in the $ct$ expression.
The two equations are equivalent; i.e. the $m_1$ and $m_2$ in Newton's gravitational equation actually represents an $m'$ where $m' = \sqrt{m_1 * m_2} = \sqrt{\varsigma_1 * \varsigma_2}$ . $m'$ is the rest mass equivalent of the two mass system located at the center of gravity between the two masses $m_1$ and $m_2$.

**Hypothesis:** *Gravity results from a volume of space flowing toward a hole-complex in space. Gravity is present (exists) without any obvious physical motion.*

The equation $E = g'ct_{g'}$ reflects the fact that gravity travels at the speed of light through time. Gravity will be considered further, following a discussion of other physical elements.



*Eric Mitchell Horn*

# Force, Momentum, and the Concepts of Other Physical Elements

These concepts are presented with their $ct$ equations in order to form a basis for the further presentation of SPT. Knowledge of these concepts is a necessary prerequisite to the understanding of the next section, Gravity Part 2.

Considering the hypothesis of scat/mass:

$$\varsigma = c^4 t_\varsigma^4$$

Applying Newton's force equation to the scat hypothesis yields the general equation for force:

$$F = ma = c^4 t_\varsigma^4 * \frac{nc}{t} = c^5 t^3$$

$$F = \frac{c}{t} * c^4 t_\varsigma^4 = c^5 t'^3$$

Or:

$$F = c^2 V_S \quad \text{where } V_S = c^3 t^3$$

Force can be conceived as a "moving, moving volume of space," Again, notice that the $n$ becomes incorporated into the final general $t^3$ element.

*The definition of the subset of **force** and the general equation for force, $F$:*

$$F = c^5 t^3$$

Maximum force:

$$F_{max} = c^5 t_\varsigma^3$$

The derived SI units for force are m²/s² x m³ = m⁵/s². Force will always have five factors of the speed of light and three factors of time.

*The definition of the subset **scat-volume** and the equation for scat-volume, $V_\varsigma$:*

$$scat\,volume = V_\varsigma = c^3 t_\varsigma^3$$

Recall that most is scat per scat-time, or the speed of light multiplied by a scat-volume:

$$most = \frac{scat}{t_\varsigma} = \frac{c^4 t_\varsigma^4}{t_\varsigma} = c^4 t_\varsigma^3 = cV_\varsigma$$

Force can be considered to be moving most:

$$force = nc * most = nc^2 V_\varsigma$$

Maximum force from one object with mass:

$$F_{max} = c^5 t_\varsigma^3 = c^2 V_\varsigma$$

The question arises whether the word "force" refers to the broad physical element of $c^5 t^3$ or whether force is a subset of a larger physical element. For example, how are force and gravity related? Is force a subset of gravity, or gravity a subset of force, or are they both independent subsets? It has already been demonstrated that gravity belongs to the same physical element as force. It can be hypothesized that many forces encountered in everyday life are "proportioned directionalized gravity." It will be shown that electromagnetic force and gravitational force are two sides of the same process; an exchange of space particles traveling at the speed of light between the inner-verse and the outer-verse. For consistency, if the name "scat" refers to the physical element of $c^4 t^4$, then the physical element of $c^5 t^3$ is accelerated scat, and the physical element is named acscat, acronym for **ac**celerated **scat**. Force and gravity will be considered to be two separate subsets of the physical element of acscat.

*The definition of the physical element of **acscat**, and the general equation for acscat:*

$$acscat = c^5 t^3$$

This categorization of acscat into two separate subsets is consistent with the categorization of the physical element of *mosp* (moving space) into the two separate subsets of *mov* and *most*.

Relating to force, the maximum accelerating distance of an object with mass:

$$d_{a'_{max}} = \frac{c}{t_\varsigma}$$

Momentum is a moving object, a moving scat/mass.

*The definition of the physical element of **momentum**, and the general equation for momentum, p:*

$$momentum = p = c^5 t^4$$

$$p = nc * c^4 t_\varsigma^4 = c^5 t^3 * t = c^5 t^4$$

The $n$ is incorporated into a general $t^4$ element.

The relationship between force, $F$, and momentum, $p$ :

$$p = Ft = c^5 t^3 * t = c^5 t^4$$

Maximum momentum:

$$p_{max} = c * c^4 t_\varsigma^4 = c^5 t_\varsigma^4$$

Momentum can be conceived as a "moving, moved volume of space."

The derived SI units for momentum are m/s x m$^4$ = m$^5$/s. In SPT, the physical element of momentum will always have five factors of the speed of light and four factors of time.

If momentum equals force multiplied by time, what is force multiplied by time squared?

$$c^5 t^3 * t^2 = c^5 t^5$$

**Hypothesis:**  *A sustained consumptive pattern is a physical element and results in scat moved a distance.*

By hypothesis, this $ct$ expression represents the physical element of a "sustained consumptive pattern." The acronym of scop, symbolized $SCP$, will be used in this text to reference this physical element. Gravity will be shown to be the result of a net continuous consumption of a volume of space from the outer-verse. A sustained force allows for the movement of matter. By hypothesis, the movement of matter through space is the result of a net sustained consumptive pattern which favors a specific directional movement. A definition is required and a general equation can be deduced.

*The definition of the physical element of **scop**, and the general equation for scop, SCP:*

$$scop = c^5 t^5$$

$$scop = SCP = c^5t^3 * t^2 = c^4t_\varsigma^4 * ct = \varsigma ct = c^5t^5$$

Scop is the product of an object with mass moved a distance. The SI units for scop are kg·m with derived SI units of a $m^5$. Since scop is also equal to the product of a force multiplied by time squared, force can be considered to be accelerating scop.

Accelerated scop equals energy:

$$\frac{c}{t} * c^5t^5 = c^6t^4 = E.$$

Energy equals force multiplied by distance:

$$E = c^5t^3 * ct_d = c^6t^4$$

$$E = c^4t_\varsigma^4 * \frac{c}{t_a} * ct_d = c^6t^4$$

$t_a$ is an acceleration time, a specific time to equal a specific accelerating distance.

*Question:* Is there a physical element of scop multiplied by the speed of light? Answer: Yes. A subset of this physical element will become very important in the further development and proof of SPT. A reasonable name for the physical element is moscop, an acronym of __mo__ving __scop.__

*The definition of the physical element of* **moscop,** *and the general equation for moscop:*

$$moscop = c^6t^5$$

$$moscop = c^5t^5 * c = c^6t^4 * t = c^6t^5$$

The professionals may be able to guess the important subset. Hint: What physical constant is representative of energy multiplied by time?

*Question:* Is there a physical element characterized by multiplying energy by time squared? Answer: Yes. Consider the equation of energy multiplied by time squared:

$$? = E * t^2 = c^6t^4 * t^2 = c^6t^6$$

*Question:* Can you guess what the $?$ is?



Eric Mitchell Horn

Answer: How about $c^6t^6 = (c^3t^3)^2 = S^2$ or "space squared." Since energy multiplied by time squared is essentially sustained energy and "space squared" can be represented by $S \times S$, a reasonable name for this physical element is the acronym of <u>s</u>ustained <u>e</u>nergy <u>s</u>pace <u>s</u>quared or sess.

*The definition of the physical element of **sess**, and the general equation for sess, $S^2$:*

$$sess = c^6t^6$$

$$sess = S^2 = c^6t^4 * t^2 = c^6t^5 * t = (c^3t^3)^2 = c^4t^4 * c^2t^2 = c^6t^6$$

The derived SI unit for the physical element of sess is a m⁶.

*Question:* What is the physical element if sess is accelerated?
Answer: Accelerated space squared, of course.

*Question:* Is there an observable physical reality associated with "accelerated space squared?"
Answer: Absolutely! This question can be answered with another question.

*Question:* Does energy (light) travel at the speed of light through time?
Answer: Yes. This is an observable physical reality. Consider this equation:

$$? = E * c * t = c^6t^4 * c * t = c^7t^5$$

Now consider this equation:

$$c^6t^6 * \frac{c}{t} = c^7t^5$$

A reasonable name for this physical element is acsess, an acronym for <u>ac</u>celerated <u>s</u>ustained <u>e</u>nergy <u>s</u>pace <u>s</u>quared.

*The definition of the physical element of **acsess**, and the general equation for acsess, $ACS$:*

$$acsess = c^7t^5$$

$$acsess = ACS = S^2 * \frac{c}{t} = c^6t^6 * \frac{c}{t} = c^6t^4 * ct = c^4t^4 * c^3t = c^6t^5 * c = c^7t^5$$

Acsess due to energy, or energy moved a distance, $ACS_E$:

$$ACS_E = Ect = c^6t^4 * ct = c^7t^5$$

The general equation of acsess due to an object with mass:

$$ACS_\varsigma = \varsigma c^3 t = c^4 t_\varsigma^4 * c^3 t = c^7 t^5$$

$ACS_\varsigma$ is the equation for a rest mass multiplied by a temperature. The proof of this appears later in this text.

Acsess due to an object with mass is scat multiplied by an accelerating volume of space. Acsess due to energy and acsess due to an object with mass are opposite processes. The concept of the physical element of acsess will prove extremely useful in the further development of gravity in the following section. The derived SI unit for acsess is a m$^7$/s$^2$.

Acsess due to energy reflects the observation that energy travels at the speed of light through time, or energy traveling a distance. An alternative perspective is to consider "**mo**ving **en**ergy," (moen):

*The definition of the physical element of **moen**, and the general equation for moen:*

$$moen = c^6t^4 * c = c^7t^4$$

Moen is power moved a distance. Power is energy/time:

$$P = \frac{energy}{time} = \frac{c^6t^4}{t} = c^6t^3$$

$$moen = P * ct = c^6t^3 * ct = c^7t^4$$

The derived SI unit for moen is a m$^7$/s$^3$.

Moen multiplied by time or "**mo**ving **en**ergy through **t**ime" equals moent:

$$c^7t^4 * t = c^7t^5$$

Moent is a subset of acsess.

   

We can now add to the previous framework:

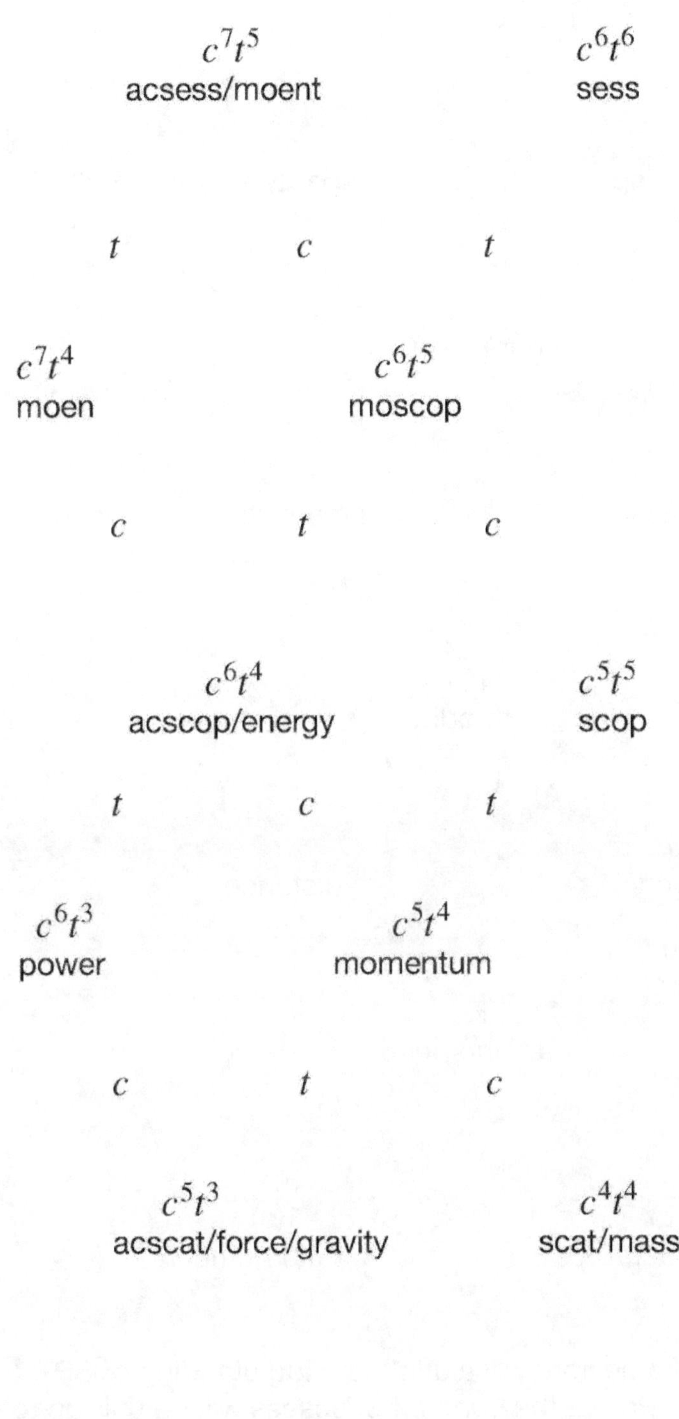

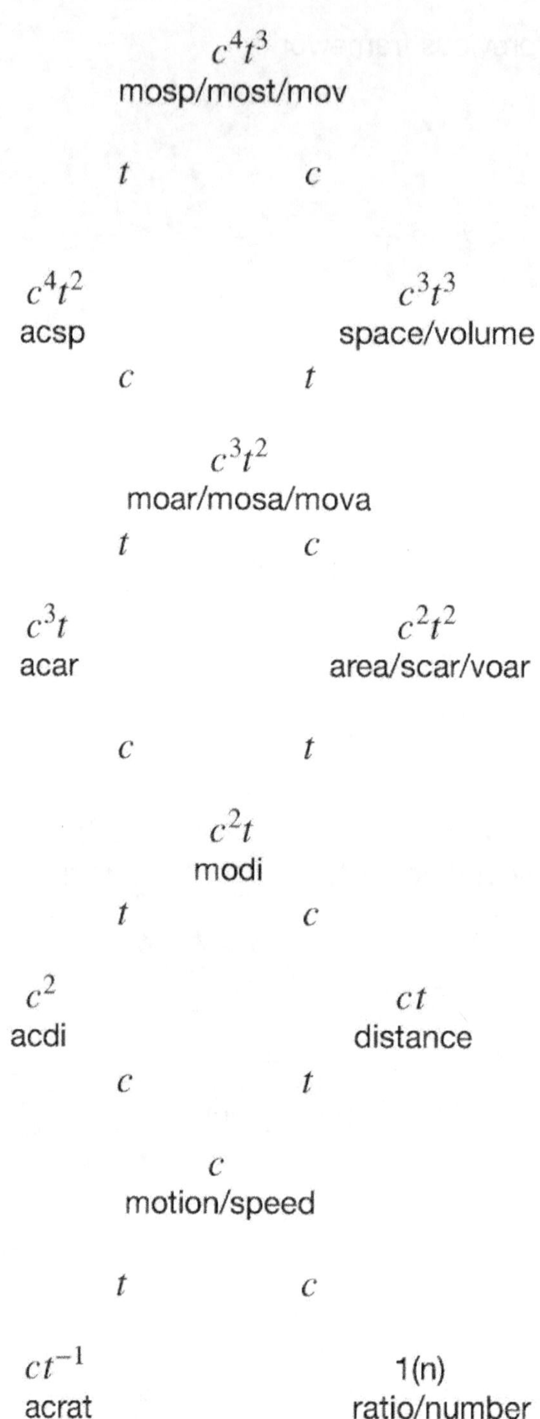


Frequent reference to this framework will aid in the understanding of SPT.
In the next chapter, Gravity: Part 2, the concept of acsess will be utilized to develop the concept of gravity further.



*Eric Mitchell Horn*

# Gravity: Part 2

After contemplating the previous hypothesis of acsess, please consider the following mathematical presentation of gravity:

$$ACS_\varsigma = \varsigma c^3 t$$

$$t(ACS_\varsigma) = \varsigma c^3 t^2$$

$$\frac{t}{c}(ACS_\varsigma) = \varsigma c^2 t^2$$

$$a = \frac{c}{t}$$

$$\frac{t}{c} = \frac{1}{a}$$

$$\frac{ACS_\varsigma}{a} = \varsigma c^2 t^2$$

$$\frac{ACS_\varsigma}{c^2 t^2} = \varsigma a = F$$

Since the expression $ct$ is a distance, $(d = ct_d)$, the units for $d^2$ $(d^2 = c^2 t_d^2)$ are contained in the $ct$ expression. The numerical value of $d$ can be separated from the $ct$ expression which carries the units; e.g. 8 m² $= 8 \times 1$m² $= 8 * c^2 t_{1m}^2$. In all of the equations, the expressions of $c$ and $t$ can be in any system of consistent units. The units are implicitly contained in the expressions of $c$ and $t$. ($t$ can be "Martian days" and $c$ can be expressed in "Jupiter diameter" per "Martian day.") For the discussion below $g''$ is the gravitational force acting between two objects with mass.

$$g''^2 = \frac{ACS_{\varsigma_1} * ACS_{\varsigma_2}}{(d^2 c^2 t^2)^2} = \frac{\varsigma_1 * c^3 t * \varsigma_2 * c^3 t}{(d^2 c^2 t^2)^2} \qquad (d^2 \text{ is unit-less})$$

$$g''^2 = \frac{ACS_{\varsigma 1}ACS_{\varsigma 2}}{(c^2 t_d^2)^2} = \frac{\varsigma_1 \varsigma_2 (c^3 t)^2}{d^4 (c^4 t_{1m}^4)}$$

$$g'' = \sqrt{\frac{\varsigma_1 * \varsigma_2}{d^4}} * \sqrt{\frac{(c^3 t)^2}{(c^2 t^2)^2}}$$

$$g'' = \frac{c^3 t}{c^2 t^2} \sqrt{\frac{\varsigma_1 * \varsigma_2}{d^4}}$$

$$g'' = \frac{c}{t_{g'}} \sqrt{\frac{\varsigma_1 * \varsigma_2}{d^4}}$$

The first variation of Newton's gravitational equation:

$$g'' = \frac{c}{t_{g'} * d^2} \sqrt{\varsigma_1 * \varsigma_2}$$

$g = c^5 t^3$ with SI derived units of m⁵s⁻² with SI derived units of $\text{m}^5\text{s}^{-2}$

This is a variation of $F = ma$. $\left(\varsigma' = \sqrt{\varsigma_1 * \varsigma_2} = \sqrt{c^4 t_{\varsigma 1}^4 * c^4 t_{\varsigma 2}^4} = c^4 t_{\varsigma}'^4\right)$

The consumptive volume, $V_c$, represents the  immediate or "instantaneous" volume of space passing through a hole complex comprising an object with mass. It is the volume of space traveling at the speed of light immediately before passing to the inner-verse. $V_{c1m^4}$ is  a net volume of space, comprised of multiple space particles, traveling through  a multi-hole-complex, simultaneously and continually traveling at the speed of light for a period of time, $t_{g'}$,  to equal one m⁴. The consumptive volume has a value much less than the scat-volume, $V_{\varsigma}$.

Recall the consumptive volume of an object with mass, $V_c$, defined:

$$V_c = (c t_V)^3$$

The relationships of scat-time, $t_\varsigma$ , volume-time, $t_V$ , and gravitational time, $t_{g'}$:

$$V_c c t_{g'} = c^4 t_\varsigma^4$$

$$c^3 t_V^3 c t_{g'} = c^4 t_\varsigma^4$$

$$t_V^3 t_{g'} = t_\varsigma^4$$

The equation for the gravitational time constant, $t_{g'}$:

$$t_{g'} = \frac{t_\varsigma^4}{t_V^3}$$

Given any object with mass expressed in m⁴, the ratio $\dfrac{t_\varsigma^4}{t_V^3} = t_{g'} = 4.4919 \times 10^{18} s$

$$t_V = \left(\frac{t_\varsigma^4}{t_{g'}}\right)^{\frac{1}{3}}$$

$$t_\varsigma = (t_V^3 t_{g'})^{\frac{1}{4}}$$

If $t_{g'}$ increases with time then gravitational acceleration may decrease with time. The precision of current methods for determining gravitational acceleration are not capable of detecting this decrease. It may be inferred that $t_{g'}$ may represent the age of the universe, 142 billion years, since by hypothesis: $matter = V_c * c * t_{g'}$. Also, $c t_{g'}$ is a distance; roughly $1.35 \times 10^{24}$ kilometers, which represents a factor for the determination of the radius of the universe. Since the space in the universe is expanding at an increasing rate, calculus is required to determine the true size of the universe. These statements warrant more Investigation.

The relationship of $k_{g'}$ to $t_\varsigma$ and $t_V$:

$$g' = \frac{\varsigma c}{t_{g'}}$$

$$k_{g'} = \frac{g'}{\varsigma}$$

$$g' t_{g'} = \varsigma c$$

$$t_{g'} = \frac{t_\varsigma^4}{t_V^3}$$

$$g' t_\varsigma^4 = \varsigma c t_V^3$$

The definition of the **gravitational acceleration constant for one scat/mass, $k_{g'}$**, and the equation for the gravitational constant for one scat/mass:

$$k_{g'} = \frac{c t_V^3}{t_\varsigma^4}$$

$k_{g'}$ reflects the fundamental relationship between the speed of light, the volume-time, and the scat-time.

The force of gravity exerted by one object with mass, $g'$:

$$g' = k_{g'} \varsigma$$

$$g' = c^5 t_V^3 = \frac{c t_V^3 * c^4 t_V^3 t_{g'}}{t_\varsigma^4} = c^5 \frac{t_V^6 t_{g'}}{t_\varsigma^4} = c^2 V_c$$

Gravity can be conceived as a "moving, moving, consumptive volume of space."

*Question:* What is the consumptive volume multiplied by the speed of light?
Answer: It is a moving volume, a subset of the physical element mosp, (moving space.)

Recall the definition of the subset $mov$ and consider the definition of gravity:

$$mov = c * c^3 t_V^3 = c^4 t_V^3 = c V_c$$

The definition of the subset of **gravity** and the equation for gravity, $g$ :

$$g = c^2 V_c$$

$$gravity = mov * c = c^5 t_V^3 = c^2 V_c$$

 Eric Mitchell Horn

Quantifying an object with mass by the scat-time allows for the quantification of relationships with other physical elements.

Equivalencies of an object with mass expressed in the derived SI unit of m⁴:

$$m_{m^4} = V_\varsigma * c * t_\varsigma = (ct_\varsigma)^3 * ct_\varsigma = c^4 t_\varsigma^4 = (c^2 t_\varsigma^2)^2 = A_\varsigma^2$$
$$= (ct_V)^3 ct_{g'} = c^4 t_V^3 t_{g'} = V_c ct_{g'}$$

Recall the scat-area of an object with mass, $A_\varsigma$, defined:

$$A_\varsigma = \sqrt{m_{m^4}} = \sqrt{V_c * c * t_{g'}} = \sqrt{c^4 t_\varsigma^4} = (ct_\varsigma)^2$$

Substituting $A_\varsigma$ for an object with mass expressed in m⁴ yields a second variation of Newton's gravitational equation:

$$g'' = \frac{c}{t_{g'} * d^2} \sqrt{A_{\varsigma 1}^2 * A_{\varsigma 2}^2}$$

$$g'' = \frac{c * A_{\varsigma 1} * A_{\varsigma 2}}{t_{g'} * d^2} \quad (A_{\varsigma 1} * A_{\varsigma 2} = \varsigma' = m_{m^4})$$

Recall that $d$ is a unit-less number; the units are contained in the $ct$ expressions. This equation is essentially the same as Newton's gravitational equation. Thus, $A_\varsigma$ is the equivalent of the gravitational mass, expressed in kilograms, in Newton's gravitational equation.  The numerical values are different. Newton's gravitational equation represents the gravity of a combined mass with the center of mass between the two objects with mass. $A_\varsigma$ is the scat-area; i.e. a moving distance multiplied by time, of the hole-complex associated with the object with mass. For this reason, Newton's gravitational equation relates a mass expressed in kilograms to an area associated with a derived mass, $m' = \varsigma$. Indeed, Newton's equation is valid only because of the relationship of $m_{m4} = A_\varsigma^2$. The units of $G_{kN}$ in Newton's equation, m³/kg·s², indicate an acceleration of a volume of space per mass, when the mass is expressed in kilograms.

A significant result of this series of equations is the determination that the "masses" of Newton's gravitational equation, which both Planck and Einstein use in their theories, do not have the same meaning as the mass in Newton's $F = ma$ or Einstein's $E = mc^2$. Considering Newton's gravitational equation, the mass considered is

$m' = \sqrt{m_1 m_2}$. Essentially: $m_{relativity} \neq m_{Newton's 2nd law} \neq m_{Newton's gravity}$

The real power of SPT lies in its ability to unmask the physical elements; to identify what the physical elements actually are. This section and the following Planck constant conversions highlight this ability.

*Conceptual Analogy:* To add some levity to this discussion, consider the following analogy. (This is a corny, but relevant analogy, which, at least in this author's mind deserves consideration as a plot in a summer blockbuster action hero movie...)

News Flash Headline: Le Monde, November 1, 2003

Dastardly Duo Nabbed in Attempted Mona Lisa Heist

Two well known criminals attempted to steal the Mona Lisa last evening. However, they were caught in the act. Jacques Cousteau, the preeminent oceanic criminologist happened to cross paths with a yellow submarine in the middle of the Atlantic. He determined that the yellow submarine was suspicious and continued to follow it. In fact, Cousteau had accidentally uncovered a plot to steal the Mona Lisa from the Louvre in the City of Light. Cousteau continued to track the pair as they travelled up the Seine in their small yellow submarine.

Luckily, and in anticipation that some drunken Halloween revelers might attempt a despoilment of the famous work of art, two preeminent crime fighters, Inspector Clouseau and his trusty assistant Cato, were dutifully tasked to guard the Mona Lisa. Cousteau radioed a timely warning to Clouseau about the arrival of the criminals.

The dastardly duo's plan was to crash the Halloween masquerade ball for the Mona Lisa's 500th birthday party. The criminals had masqueraded themselves as Joker and Penguin of Batman infamy. In fact, the criminals were Joker and Penguin. Their disguise should have worked. The criminals drugged the revelers in the adjoining room with laughing gas, then proceeded to pass through a secret underground tunnel and subsequently up through a secret trap door in the floor of Mona Lisa's display room.

Inspector Clouseau, having been forewarned, installed a $2\pi$ radian surround retinal scanner to confirm the thieves identities as they came through the trap door. Caught in the act, Joker and Penguin pleaded innocent in French court today because they couldn't take off their masks, since their masks were superglued to their faces. Thus mask removal would require significant pain and suffering and involve multiple facial plastic surgeries. Their defense apparently revolved around this cruel and unusual punishment.

However, in French court today, Inspector Clouseau, was able to unmask the masked criminals and prove Joker and Penguin were, in fact, who they were masquerading as



*Eric Mitchell Horn*

by virtue of their retina scans.  The court agreed with Inspector Clouseau. Joker and Penguin were deemed to be two and the same, and were sentenced to life behind bars for eternity, with no access to lollipops.

Now, returning to the subject at hand, after this brief, but hopefully entertaining interlude:

Substituting $(ct_\varsigma)^2$ for $A_\varsigma$ in the above expression for gravity yields a third variation of Newton's gravitational equation in terms of $c$, $t_\varsigma$, and $t_{g'}$ and the unit-less variable $d$. $t_{\varsigma1}$ and $t_{\varsigma2}$ are times specific to a scat-length, $ct_\varsigma$ , such that, $(ct_\varsigma)^2$ equals the scat-area, $A_\varsigma$ .

$$g'' = \frac{c}{t_{g'} * d^2} * (ct_{\varsigma1})^2 * (ct_{\varsigma2})^2$$

$$g'' = \frac{c^5 * t_{\varsigma1}^2 * t_{\varsigma2}^2}{t_{g'} * d^2}$$

Expressing the objects with mass by components yields a fourth variation of Newton's gravitational equation:

$$g'' = \sqrt{\frac{V_{c_1} * c * t * c^3 t * V_{c_2} * c * t * c^3 t}{(d^2 c^2 t^2)^2}} \quad = \sqrt{\frac{V_{c_1}^2 * V_{c_2}^2 * (ct)^2 * (c^3 t)^2}{(dct)^4}}$$

$$V_c = \frac{m_{m^4}}{ct_{g'}} = \frac{c^4 t_\varsigma^4}{ct_{g'}}$$

$$g'' = \frac{c^2}{d^2}\sqrt{V_{c_1} * V_{c_2}} \quad (\sqrt{V_{c_1} * V_{c_2}} = V_c' \; ; d^2 \text{ is unit-less})$$

*Question:* What is the value of $k_{g''}$, the gravitational constant for the force of gravitational attraction between two objects with mass, expressed in m⁴, a distance $ct_d$ (in meters) apart in this system of units?
Answer: $k_{g''} = 2.9979 \times 10^8$ m⁻¹s⁻² See the derivation below.

Assume the numerical value of $k_{g''} = G_{kN} * t_{g'}$ . Performing $ct$ analysis and incorporating $t_{g'}$ into general $t$:

$$c^5 t^3 = \frac{1}{c^3 t^5} * c^4 t^4 * c^4 t^4 = \frac{c^8 t^8}{c^3 t^5} = \frac{c^8 t^8}{c t^3 * (c^2 t_d^2)}$$

$k_{g''}$ must have $ct$ units and expression of $c^{-1} t^{-3}$ or SI derived units of m⁻¹s⁻² with a numerical value equal to:

$$k_{g'} * t_{g'} = \#c.$$

$$k_{g''} = k_{g'} * t_{g'} = 6.6741 \times 10^{-11} * 4.4919 \times 10^{18} = 2.9979 \times 10^8 \text{ m}^{-1}\text{s}^{-2}$$

$$k_{g''} = 2.9979 \times 10^8 \text{ m}^{-1}\text{s}^{-2}$$

Before confirming this with an example, the conversion factor between an object with mass expressed in kilograms and an object with mass expressed in m⁴ ($m_{m4}$) must be determined.

# Conversion of Kilogram to Meter⁴ and Gravity: Part 3

What is the conversion between a kilogram and a m⁴? Let the value of $m_1 = m_2 = d^2$ $= 1$ kg and 1 m, then $V_{c1} = V_{c2}$ and $g$ is equal to the numerical value, $F_g$, the gravitational force in newtons.  For the following derivation, $G_N$ represents the gravitational force in newtons of one object with mass. However, it can be used to determine a conversion factor between a kilogram and a m⁴.  $V_c$ is the consumptive volume of a mass of 1 m⁴.

$$G_{kN} * \frac{m_{1kg} * m_{1kg}}{d^2} = F_g = 6.674 \times 10^{-11} \text{ N} = \sqrt{\frac{ACS_{\varsigma 1} * ACS_{\varsigma 2}}{(c^2 t_d^2)^2}}$$

$$G_N = c^2 \sqrt{V_c^2}$$

$$\frac{G_N}{c^2} = V_c$$

This yields a general equation for the gravitational force of one mass:

$$g' = c^2 V_c = c^5 t_V^3$$

Recall:  $$V_c = \frac{m_{m^4}}{c t_{g'}} = \frac{\varsigma}{c t_{g'}} \; ; \; t_{g'} = 4.4919 \times 10^{18} \text{s}$$

The consumptive volume of 1 m⁴:

$$V_{c1m^4} = \frac{1}{c t_{g'}} = 7.4259 \times 10^{-28} \text{ m}^3 \qquad \text{(The "1" in this equation has units of m⁴)}$$

Physically, an object with mass is an object with mass regardless of how the mass is defined or the units used. In this context, a kg/m represents a volume of space. A mass expressed in kilograms is the same as a mass expressed in m⁴. (The numerical values are not equal.)

$$\frac{kg}{ct} = k * \frac{(ct)^4}{ct}$$ where  $k$  is a conversion factor, a time ratio, between the two units of mass.

*The definition of a mass of 1 kilogram in m⁴, t=1 second:*

$$m_{m^41kg} = 7.4259 \times 10^{-28} \text{m}^3 \text{ x } c \text{ x } t = 2.2262 \times 10^{-19} \text{ m}^4$$

*The mass conversion constant:*

$$2.2262 \times 10^{-19} \text{ m}^4/\text{kg}$$

This constant reflects the ratio of time-lengths between a kilogram and a mass expressed in m⁴.

$$1 \text{ kg} = 2.2262 \times 10^{-19} \text{m}^4 = c^4 t^4_{\varsigma 1kg} = \text{ when } t_{\varsigma 1kg} = 7.2456 \times 10^{-14}\text{s}$$

$t_{\varsigma 1kg}$ is the 1 kilogram scat-time.

$$1 \text{ m}^4 = c^4 t^4 \text{ when } t_{\varsigma 1m^4} = 3.3356 \times 10^{-9}\text{s}$$

$3.3356 \times 10^{-9}$s is the scat-time of one m⁴.

The scat-time ratio, $k_{t^4_{\varsigma 1kg} t^4_{\varsigma 1m^4}}$:

$$k_{t^4_{\varsigma 1kg} t^4_{\varsigma 1m^4}} = \frac{t^4_{\varsigma 1kg}}{t^4_{\varsigma 1m^4}} = \frac{(7.2459 \times 10^{-14})^4}{(3.3356 \times 10^{-9})^4} = 2.2262 \times 10^{-19} \text{ (unit-less)}$$

To confirm the previous hypotheses for mass and energy are consistent with known values and observation, consider the electron rest mass energy:

Checking, given the rest mass energy of the electron is 0.511 MeV. Expressing this energy in units of m⁶s⁻² yields:

$$5.11 \times 10^5 \text{eV} * 1.6022 \times 10^{-19} \text{ J/eV} * 2.2262 \times 10^{-19} \text{ m}^4\text{kg}^{-1} = 1.8227 \times 10^{-32} \text{ m}^6\text{s}^{-2}$$

$$\varsigma_e = m_e * 2.2262 \times 10^{-19} = 2.0280 \times 10^{-49}\text{m}^4$$

$$A_{\varsigma e} = \sqrt{\varsigma_e} = 4.5033 \times 10^{-25}\text{m}^2$$

$$E_e = (cA_{\varsigma e})^2 = (c * 4.5033 \times 10^{-25})^2 = 1.8226 \times 10^{-32}\text{m}^6\text{s}^{-2}$$



*Eric Mitchell Horn*

The obtained energy values are equivalent within rounding errors. This confirms the relative relationship between a mass expressed in kilograms and a scat equivalent expressed in m⁴.

**Hypothesis:** *Any conversion constant between any physical element expressed in SI or any other system of units and the same physical element expressed $ct$ units will be a unit-less ratio of powers of the respective time-lengths.*

Many relationships between constants involving different physical elements will relate powers of $c$, $t_\varsigma$, and $t_V$. and other time constants. Determining the conversion factor between a mass expressed in kilograms and a mass expressed in m⁴, and knowing that gravity must have units of m⁵/s² must mean that $t^2$ is seconds squared which results in one unknown $t$ since $t_{1mtr} = 3.3356 \times 10^{-9}$s; allowing for confirmation by example. Incorporating $d^2$ into the general equation, and using the specific masses of 1kg and 2kg allows for a calculation of $k_{g''}$:

$$g'' = \frac{1}{d^2 c^2 t_{1mtr}^2 * ct * t^2} * c^4 t_{\varsigma 1kg}^4 * c^4 t_{\varsigma 2kg}^4$$

$$F_g = G_{kN} \frac{m_{kg1} m_{kg2}}{d^2}$$

Let $d = 1$ m, $m_1 = 1$ kg, $m_2 = 2$ kg

$$F_g = 6.674 \times 10^{-11} * 1 * 2 = 1.3348 \times 10^{-10} \text{ N}$$

$$1.3348 \times 10^{-10} * 2.2262 \times 10^{-19} = 2.9716 \times 10^{-29} \text{ m⁵/s²}$$

$$g'' = 2.9716 \times 10^{-29} = \frac{2.2262 \times 10^{-19} * 4.4525 \times 10^{-19}}{1 * c^2 * (3.3356 \times 10^{-9})^2 * ct * s^2}$$

Solving for $t$:  Time $= t = 1.1127 \times 10^{-17}$s $= \#\dfrac{1}{c^2}$ (s)

$$k_{g''} = 2.9979 \times 10^8 \text{m}^{-1}\text{s}^{-2}$$

Checking:

$$2.9716 \times 10^{-29} = 2.9979 \times 10^8 * 2.2262 \times 10^{-19} * 4.4525 \times 10^{-19}$$

A fifth variation of Newton's gravitational  equation:

$g'' = k_{g''}\dfrac{\varsigma_1\varsigma_2}{d^2}$ when the objects with mass are expressed in m⁴ and $k_{g''}$ has units of m⁻¹s⁻² with a $ct$ expression of $c^{-1}t^{-3}$ and $d^2 = c^2t_d^2$ and has units of m².

$$g'' = k_{g''} * \frac{c^4 t_{\varsigma 1}^4 c^4 t_{\varsigma 2}^4}{c^2 t_d^2}$$

$$g'' = 2.9979 \times 10^8 * \frac{\varsigma_1\varsigma_2}{d^2}$$

$$k_{g''} = 2.9979 \times 10^8 = c^{-1}t^{-3} = \frac{1}{c\,t_{g''}^3}$$

$$t_{g''} = 2.2325 \times 10^{-6}\text{s}$$

$$g'' = k_{g''} * \frac{c^4 t_{\varsigma 1}^4 c^4 t_{\varsigma 2}^4}{c^2 t_d^2} = \frac{c^4 t_{\varsigma 1}^4 c^4 t_{\varsigma 2}^4}{c^3 t_{g''}^3 t_d^2}$$

Reconsidering the $ct$ expression for the gravitational force acting between two masses:

$$c^5 t^3 = \frac{c^4 t^4 * c^4 t^4}{c^3 t^5}$$

Although more complicated, a sixth variation of Newton's gravitational equation:

$$g'' = c^5 t^3 = \frac{c^4 t_{\varsigma 1}^4 * c^4 t_{\varsigma 2}^4}{c^3 t^3 * t_d^2} = \frac{\varsigma_1\varsigma_2}{V_S * t_d^2} \text{ where } V_S \text{ has the numerical value of } \frac{d^2}{c}$$

Incorporating constants:

$$V_S = \frac{4\pi r^3}{3} = \frac{d^2}{c} \text{ (m}^3\text{)} \qquad (r^3 = \frac{3d^2}{4\pi c} ; r = 9.2689 \times 10^{-4} * d^{\frac{2}{3}} \text{ (}r\text{ is expressed in meters)}$$

Considering time-lengths; the gravitational force acting between two objects with mass:

$$k_{g''} = \frac{1}{c\,t_{g''}^3}$$



*Eric Mitchell Horn*

$$g'' = c^5 t_{g''}'^3 = k_{g''} \frac{\varsigma_1 \varsigma_2}{c^2 t_d^2} = \frac{c^4 t_{V_1}^3 t_{g'} * c^4 t_{V_2}^3 t_{g'}}{c t_{g''}^3 * c^2 t_d^2}$$

$$t_{g''}'^3 = \frac{t_{V_1}^3 t_{V_2}^3 t_{g'}^2}{t_{g''}^3 t_d^2}$$

$\dfrac{t_{g'}^2}{t_{g''}^3}$ is an inverse time constant, $k_{t_g^2 t_g^3{}_{''}}$:

$$k_{t_g^2 t_g^3{}_{''}} = t^{-1} = \frac{t_{g'}^2}{t_{g''}^3} = \frac{2.0177 \times 10^{37}}{1.1127 \times 10^{-17}} = 1.8134 \times 10^{54} \text{s}^{-1} = \frac{1}{5.5146 \times 10^{-55} s}$$

$$t_{g''}'^3 = k_{t_g^2 t_g^3{}_{''}} * \frac{t_{V1}^3 t_{V2}^3}{t_d^2}$$

$$c^5 k_{t_g^2 t_g^3{}_{''}} = 4.3913 \times 10^{96} \text{ m}^5\text{s}^{-6}$$

A seventh variation of Newton's gravitational equation:

$$g'' = 4.3913 \times 10^{96} * \frac{t_{V_1}^3 t_{V_2}^3}{t_d^2} \quad \text{(m}^5\text{/s}^2)$$

Considering the relationship between $F_{max}$ and $g$:

$$F_{max} = c^5 t_\varsigma^3$$

$$g' = c^2 V_c = c^5 t_V^3$$

$$\frac{F_{max}}{g'} = \frac{c^5 t_\varsigma^3}{c^5 t_V^3} = \frac{t_\varsigma^3}{t_V^3}$$

$$F_{max} = \frac{g' t_\varsigma^3}{t_V^3}$$

$$t_\varsigma F_{max} = g' t_{g'}$$

$$F_{max} = \frac{g' t_{g'}}{t_\varsigma}$$

$$t_{g'} = \frac{t_\varsigma^4}{t_V^3} = \frac{F_{max} t_\varsigma}{g'}$$

$$t_\varsigma^3 = \frac{t_V^3 F_{max}}{g'}$$

$$g' = \frac{t_V^3 F_{max}}{t_\varsigma^3}$$

Gravity and the relationships of $g'$ to $g''$ and $k_{g'}$ to $k_{g''}$ will be considered further, following a discussion of charge. SPT's real power is demonstrated in the following section regarding Planck units and their relationship with SPT.

# Comparison and Conversion of Planck Units

In this context, Planck's $G$, [Newton's $G$] is equivalent to $k_{g''}$, the constant that relates the gravitational attraction between two masses expressed in m⁴. This section, presented following the discussion gravity and the true relationship between two scat/masses, will serve to support later hypotheses for charge, temperature, entropy, and the general equation for scat/mass. For specific conversions, see the section "Selected Constant Conversions and Relationships." Planck mass in kg x 2.2262 x10⁻¹⁹ ÷ $c^2$ numerically equals Planck time.

Because Planck used Newton's gravitational constant, which was assumed to relate two masses, these values may differ from Planck's; i.e. Newton's $G$ does not numerically equal $k_{g''}$. The important finding of note is that Planck's units, modified using the correct gravitational constant between two rest masses (scat quantities), match the derived $ct$ units in SPT.

$$h'_b = 2.3477 \times 10^{-53} \text{m}^6/\text{s}$$

$$k_{g''} = 2.9979 \times 10^8 \text{m}^{-1}\text{s}^2$$

$$k'_B = 1.5812 \times 10^{-41} \text{m}^3$$

$$\varepsilon'_0 = 7.8848 \times 10^{-44} \text{m}$$

By $ct$ analysis:

Planck time:

$$t'_P = \sqrt{\frac{h'_b k_{g''}}{c^5}} = \sqrt{\frac{c^6 t^5 * c^{-1} t^{-3}}{c^5}} = \sqrt{t^2} = t$$

when $t_P = 5.3911 \times 10^{-44}\text{s}$

Planck length:

$$l'_P = \sqrt{\frac{h'_b k_{g''}}{c^3}} = \sqrt{\frac{c^6 t^5 * c^{-1} t^{-3}}{c^3}} = \sqrt{c^2 t^2} = 1.6162 \times 10^{-35} \text{ m} = ct$$

when $t_{l_P} = 5.3911 \times 10^{-44}\text{s}$

Planck temperature:

$$T_P' = \sqrt{\frac{c^5 h_b'}{k_{g''}(k_B')^2}} = \sqrt{\frac{c^5 * c^6 t^5}{c^{-1}t^{-3} * (c^3 t^3)^2}} = \sqrt{c^6 t^2} = 2.7541 \times 10^{31}\text{m}^3\text{/s}^2 = c^3 t$$

when $t_{T_p} = 1.0221 \times 10^6$s

Note: This value is not a true maximum temperature. The SPT units are correct, but Boltzmann's constant is not a true quantum constant since a volt is not a quantum moving distance. The true maximum theoretical temperature is given by $c^3 t_{\varsigma_P}$ which will be derived in a later chapter.

Planck entropy (derived from Planck temperature):

$$S_P' = \sqrt{\frac{c^5 h_b'}{k_{g''}T_P'^2}} = \sqrt{\frac{c^5 * c^6 t^5}{c^{-1}t^{-3} * (c^3 t)^2}} = \sqrt{c^6 t^6} = 1.5812 \times 10^{-41}\text{m}^3 = c^3 t^3$$

when $t_{S_p} = 8.3722 \times 10^{-23}$s

Planck charge:

$$Q_p' = \sqrt{4\pi \varepsilon_0' h_b' c} = \sqrt{4\pi * ct * c^6 t^5 * c} = \sqrt{c^8 t^6} = 8.3509 \times 10^{-44}\text{m}^4\text{/s} = c^4 t^3$$

when $t_{Q_P} = 2.1785 \times 10^{-26}$s

Planck mass:

$$m_P' = \sqrt{\frac{h_b' c}{k_{g''}}} = \sqrt{\frac{c^6 t^5 * c}{c^{-1}t^{-3}}} = \sqrt{c^8 t^8} = 4.8453 \times 10^{-27}\text{m}^4 = c^4 t^4$$

when $t_{\varsigma_P} = 8.8005 \times 10^{-16}$s

The above $ct$ expressions will be used to confirm further hypotheses concerning charge, temperature, and entropy presented in the following sections. Although this chapter is presented in this book before many of the physical elements and their subsets, the conversion of these Planck units occurred following SPT's development of charge, temperature and entropy.

# Charge and Related Physical Elements

*Conceptual analogy:* Imagine that you are on the Greenland glacier in January. Put on some gloves because it is freezing cold outside. Consider your gloves.

*Question:* Does your left handed glove fit your right hand?
Answer: No. The difference is a result of a property of three dimensional space, known as chirality. The right hand is the mirror image of the left hand.

*Conceptual analogy:* Perform a mind experiment envisioning two tornadoes. Imagine that you are in a helium balloon viewing the tornadoes from above. One of the tornadoes is rotating clockwise, the other counter-clockwise.

*Question:* Do these opposite directionalized air flows result in the tornados moving toward one another or away from one another?
Answer: The opposite directionalized air flows will move toward one another. To confirm, draw these imaginary air flows on a piece of paper.

*Question:* If the tornadoes have air flows moving the same direction; i.e. two clockwise tornadoes or two counterclockwise tornadoes, do they move toward one another or away from one another?
Answer: The tornadoes will move away from one another. Confirm by drawing the air flows on a piece of paper.

*Question:* How does this analogy apply to physical phenomenon in SPT? (Opposite airflows attract. Similar airflows repel.)
Answer: Electrons and positrons, matter and antimatter, could be opposite 3D space particle flows.

**Hypothesis:** *Two like charges repel one another because of the chirality involved in the individual space particle flows. Two unlike charges attract one another for the same reason.*

**Hypothesis:** *All negatively charged "particles"; e.g. electrons, muons, tau particles, down quarks, strange quarks, bottom quarks, repel one another because of similar directionalized space particle flows. All positive charged "particles"; e.g. protons, positrons, up quarks, top quarks, charm quarks repel one another because of similar directionalized space particle flows.*

This theory proposes that the observed property of charge results from a net directionalized consumption of space particles. Opposite charges have an opposing chirality of consumption. Considering the electron as a hole-complex in space, the specific geometry of the hole-complex of the electron is unknown. However, it can be surmised that the resulting space particle flows create the observed property of charge.

*Question:* What are the physical meanings of the known relationships of charge, current and force?
Answer: A magnetic field is a force field; i.e. a magnetic field, abbreviated "$m.f.$", like any force, will have the same $ct$ expression as any other force.

$$m.f. = c^5t^3 = c*?$$

$$? = c^4t^3$$

It is known that a moving electric charge creates a magnetic field. Therefore, charge, symbolized $Q$, must equate with the question mark. This simple observation that moving charge generates a force yields the following hypothesis.

$$F = c^5t^3 = c*c^4t^3 = cQ$$

**Hypothesis:** *Charge is a directionalized moving consumption of a volume of space at the speed of light.*

$$Q = c^4t^3$$

*Question:* What are the derived SI units of charge if a mass is considered to be a $m^4$?
Answer: $m^4/s^4 \times s^3 = m^4/s$

In this text, the symbol $Q$ will denote charge in general. The symbol $Q_e$ will be used to denote electron charge specifically. This will avoid confusion with the symbol $e$ which may be used to reference the electron. Traditionally, the symbol $e$ represents the quantum charge of the electron.

Conversion of $e$ to $Q_e$ in units of $m^4/s$ to equal $c^4t^3$:

$$e = 1.6022 \times 10^{-19}\,C = 1.6022 \times 10^{-19}\,A{\cdot}s$$
$1A = 4.4525 \times 10^{-26}\,m^4s^{-2}$ (See Appendix B, SI unit to $ct$ conversions.)

$$e' = Q_e = (1.6022 \times 10^{-19})(4.4525 \times 10^{-26}) = 7.1338 \times 10^{-45}m^4/s$$

$$Q_e = 7.1338 \times 10^{-45}\,m^4/s = c^4t^3 \text{ when } t = 9.5943 \times 10^{-27}s$$

*Question:* Does the $t$ in electron charge equation relate to a volume-time or a scat-time?

Answer: The observation that force is directionalized requires a time element that relates to the scat-time and not the volume-time, since a direction of movement cannot be detected or perceived without a time element. Also consider the following:

Conversion of electron rest mass to electron scat:

$$m_e = 9.1094 \times 10^{-31} \text{kg}$$

$$\varsigma_e = 9.1904 \times 10^{-31} * 2.2262 \times 10^{-19} = 2.0460 \times 10^{-49} \text{m}^4$$

$$\varsigma_{Q_e} = Q_e * t_{g'} = 7.1338 * 10^{-45} * 4.4919 \times 10^{18} = 3.2044 \times 10^{-26} \text{m}^4$$

If the $t$ of electron charge were a volume-time, the scat equivalent of electron charge would be much greater than the scat equivalent of the electron which eliminates the possibility that this time is a volume-time. Therefore, the $t$ of electron charge must be a scat-time.

$$Q_e = cV_{\varsigma Q_e} = c(c^3 t^3_{\varsigma Q_e}) = c^4 t^3_{\varsigma Q_e}$$

The volume of space consumed in one scat-time of the charge of the electron; i.e. the electron charge scat-volume, $V_{\varsigma Q_e}$:

$$V_{\varsigma Q_e} = \frac{Q_e}{c} = c^3 t^3_{\varsigma Q_e} = \frac{7.1338 \times 10^{-45}}{c} = 2.3796 \times 10^{-53} \text{m}^3$$

Charge is a subset of most which itself is a subset of the physical element mosp.

*The definition of the subset of* **charge**, *and the ct equation for charge, Q:*

$$Q = c^4 t^3_{\varsigma Q}$$

*The definition of the* **electron charge quantum** *and the numerical value of the electron charge quantum,* $Q_e$:

$$Q_e = c^4 t^3_{\varsigma Q_e} = 7.1338 \times 10^{-45} \text{m}^4/\text{s}$$

*The definition of the **electron charge scat equivalent**, $\varsigma_{Q_e}$:*

$$\varsigma_{Q_e} = c^4 t^4_{\varsigma_{Q_e}}$$

*The definition of **electron charge scat-time** and the numerical value of electron charge scat-time, $t_{\varsigma_{Q_e}}$:*

$$t_{\varsigma_{Q_e}} = (\frac{\varsigma_{Q_e}}{c^4})^{\frac{1}{4}} = 9.5943 \times 10^{-27} \text{s}$$

*Question:* How does electron charge scat compare with electron scat?
Answer: The ratio of electron scat, $\varsigma_e$, to the electron charge scat equivalent, $\varsigma_{Q_e}$:

$$\varsigma_{Q_e} = c^4 t^4_{\varsigma_{Q_e}} = 6.8445 \times 10^{-71} \text{m}^4 = 3.0744 \times 10^{-52} \text{kg}$$

$$t^4_{\varsigma_e} = 2.5106 \times 10^{-83} \text{s}^4$$

$$t^4_{\varsigma_{Q_e}} = 8.4733 \times 10^{-105} \text{s}^4$$

$\varsigma_e$ to $\varsigma_{Q_e}$ ratio:

$$\frac{m_e}{m_{Q_e}} = \frac{9.1094 \times 10^{-31}}{3.0744 \times 10^{-52}} = \frac{t^4_{\varsigma_e}}{t^4_{\varsigma_{Q_e}}} = \frac{2.5106 \times 10^{-83}}{8.4733 \times 10^{-105}} = 2.9630 \times 10^{21} \text{ (unit-less)}$$

The electron charge is a very small component of the electron as a whole.

The definition of charge allows for development of other physical elements related to charge. But first consider the physical element of accelerated space, with an acronym acsp for **ac**celerated **sp**ace. An accelerated space is also accelerating scat.

*The definition of the physical element **acsp**, and the general $ct$ equation for acsp:*

$$acsp = c^4 t^2$$

$$acsp = \frac{c^4 t^3}{t} = \frac{c^5 t^3}{ct} = \frac{c^4 t^4}{t^2} = c^3 t * ct = \frac{c}{t} * c^3 t^3 = c^4 t^2$$

*Question:* How do charge and current relate?



Eric Mitchell Horn

Answer: Current, symbolized $I$, is charge per time, a subset of acsp.

$$current = \frac{charge}{time} = \frac{Q}{t} = \frac{c^4 t^3}{t} = c^4 t^2$$

Since charge involves a scat-time, current will also involve a scat-time.

*The definition of the subset of **current**, and the $ct$ equation for current, $I$:*

$$I = c^4 t^2_{\varsigma_I}$$

*Question:* What are the derived SI units for current?
Answer: Current has SI derived units of m⁴/s².

Coulomb's law relates the force exerted between two charges. By Coulomb's law:

$$F = k_e \frac{Q_1 * Q_2}{d^2}$$

Translating units of $k_e$ to an expression in terms of $c$ and $t$:

$$k_e = \text{N•m²/C²} = \frac{c^7 t^5}{c^8 t^6} = \frac{1}{ct}$$

The $ct$ expression analysis for Coulomb's Law:

$$F = c^5 t^3 = \frac{1}{ct} * \frac{c^4 t_1^3 * c^4 t_2^3}{c^2 t^2} = \frac{c^6 t^4}{ct} = \frac{E}{ct} = \frac{E}{d} \qquad \text{(A variation of } E = Fd)$$

An equation for determining the force between two charges:

$$F''_Q = \frac{1}{ct} * \frac{Q_1 Q_2}{(ct_d)^2} = \frac{Q_1 Q_2}{c^3 t'^3}$$

$(d^2 = c^2 t_d^2$; the units for $d$ are contained in the $ct$ expression.)

$$\frac{Q}{ct} = \frac{c^4 t^3}{ct} = c^3 t^2 = \frac{V_S}{t} = \sqrt{E_Q}$$

$$V_\varsigma = \frac{Q}{c}$$

$$Q = cV_\varsigma$$

$$e' = Q_e = ct_{\varsigma Q_e}\sqrt{E_{Q_e}} = ct_{\varsigma Q_e}\sqrt{V_{c_{Q_e}}c^3 t_g} = ct_{\varsigma Q_e}\sqrt{c^3 t_{V_{Q_e}}^3 \, c^3 \frac{t_{\varsigma Q_e}^4}{t_{V_{Q_e}}^3}} = c^4 t_{\varsigma Q_e}^3$$

$t_{\varsigma Q_e}^3$ is electron charge scat-time cubed. $V_{\varsigma Q_e}$ is the electron charge scat-volume; i.e. the electron charge space, $S_{Q_e}$. The electron charge scat-volume means the same as the electron charge space. This will become apparent after later discussion.

*The definition of **electron charge space**, and the $ct$ equation for electron charge space, $S_{Q_e}$:*

$$S_{Q_e} = V_{\varsigma Q_e} = c^3 t_{\varsigma Q_e}^3$$

Checking for the force acting between two electrons 1 meter apart:

$$F_N = (\frac{c^2}{10^7})\frac{Q_1 Q_2}{d^2} = (\frac{c^2}{10^7})(\frac{(1.6022 \times 10^{-19})^2}{1^2}) = 2.3071 \times 10^{-28}\text{N}$$

$$\times\, 2.2262 \times 10^{-19} = 5.1362 \times 10^{-47} \text{ m}^5/\text{s}^2$$

Since $k_e$ relates the force acting between two charges, the notation for $k_e$ as expressed in SI derived units will be $k_{e''}$. The force acting between two charges will be denoted by $F_Q''$.

$k_e$ has units of kg·m³/C²·s² or kg·m³/A²·s⁴ with derived units of m⁻¹

$$k_{e''} = \frac{(8.9876 \times 10^{-9})(2.2262 \times 10^{-19})}{(4.4525 \times 10^{-26})^2} \text{ m}^{-1}$$

$$k_{e''} = 1.0093 \times 10^{42} \text{ m}^{-1}$$

$$1.0093 \times 10^{42} = c^{-1}t^{-1} \text{ when } t = 3.3050 \times 10^{-51}\text{s}$$



*Eric Mitchell Horn*

$$k_{e''} = \frac{1}{ct_{k_e}} = \frac{1}{4\pi\varepsilon_0} = \frac{1}{4\pi ct_f}$$      (See page 127. $t_f$ is a fundamental time constant.)

$$t_{k_e} = 4\pi t_f$$

The first variation of Coulomb's law:

$$F''_Q = k_{e''} * \frac{Q_1 Q_2}{c^2 t_d^2}$$

$$F''_{Q_e} = k_{e''} * \frac{Q_e^2}{1^2} = 1.0093 \times 10^{42} * \frac{(7.1338 \times 10^{-45})^2}{1^2} = 5.1364 \times 10^{-47} \text{m}^5/\text{s}^2$$

Within rounding error, this value confirms the numerical value of charge in $ct$ compatible units.

Considering the $ct$ analysis for force between two charges a distance apart:

$$c^5 t^3 = \frac{c^4 t^3 * c^4 t^3}{c^3 t^3} = \frac{c^8 t^6}{c^3 t^3} = k_{e''} * \frac{c^4 t^3 * c^4 t^3}{c^2 t_d^2}$$

$$F''_Q = \frac{1}{4\pi ct_f} * \frac{Q_1 Q_2}{(ct_d)^2} = \frac{Q_1 Q_2}{c^3 t^3} \qquad (t^3 = 4\pi t_f t_d^2 = t_d^2 t_{k_e})$$

Coulomb's law can be translated to $F = \dfrac{k_Q Q^2}{V_S}$ where $k_Q$ is the unit-less number 1.00926 x 10⁴², and $V_S$ is a volume of space equal to the numerical value of $d^2$, Incorporating the constant $k_Q$ into $V_S$ yields:

$$F''_Q = \frac{k_Q Q_1 Q_2}{V_S} \text{ where } V_S = \frac{4}{3}\pi r^3 \text{ and } r = 0.23873 d^{\frac{2}{3}}$$

A second variation of Coulomb's law with $k_{e''}$ incorporated into the equation:

$$F''_Q = c^5 t^3 \text{ when } t'^3 = \frac{t_{Q_1}^3 t_{Q_2}^3}{4\pi t_f t_d^2}$$

$$F''_Q = \frac{Q_1 Q_2}{c^3 t_{k_e} t_d^2}$$

$\dfrac{c^5}{t_{k_e}}$ is a constant $= 7.3271 \times 10^{92} \, \text{m}^5/\text{s}^6$

Combining constants yields a third variation of Coulomb's law:

$$F''_Q = \frac{Q_1 Q_2}{c^3 t_{k_e} t_d^2} = \frac{c^5 t_{\varsigma Q_1}^3 t_{\varsigma Q_2}^3}{t_{k_e} t_d^2} = 7.3271 \times 10^{92} * \frac{t_{\varsigma Q_1}^3 t_{\varsigma Q_2}^3}{t_d^2} \quad (\text{m}^5/\text{s}^2)$$

$k_{e''}$ is part of the $c^3 t^3$ expression.

The equation for a volume of a sphere:

$$V_S = \frac{4\pi r^3}{3}$$

Relating this spherical volume to a cubic volume:

$$V_S = c^3 t^3 = \frac{4}{3}\pi r^3$$
$$k_{e''}^{-1} = 4\pi c t_f$$

$$\frac{4\pi r^3}{3} = 4\pi c t_f c^2 t_d^2$$

$$\frac{r^3}{3} = c t_f c^2 t_d^2$$

$$r = (3 c t_f c^2 t_d^2)^{\frac{1}{3}}$$

$$V_S = c^3 t^3 = \frac{4\pi [(3 c t_f c^2 t_d^2)^{\frac{1}{3}}]^3}{3} = 4\pi c^3 t_f t_d^2 = 4\pi c t_f c^2 t_d^2$$

This demonstrates that the factor of $4\pi$ relates a cubic volume to a spherical volume.

$$F_Q'' = c^5 t'^3 = \frac{Q_1 Q_2}{4\pi c^3 t_d^2 t_f} = \frac{c^5 t_{\varsigma Q_1}^3 t_{\varsigma Q_2}^3}{4\pi t_d^2 t_f}$$

With units of m⁴/s, other physical elements are related to charge: a scat/mass per time, a force divided by the speed of light, and energy divided by voltage.

$$Q = cV_\varsigma = \frac{\varsigma}{t} = \frac{F}{c} = \frac{E}{V_{ltg}}$$

Voltage, symbolized $V_{ltg}$ (disambiguation from volume, symbolized $V$), can be assigned a $ct$ expression.

$$V_{ltg} = \frac{E}{Q} = \frac{c^6 t_{\varsigma Q}^4}{c^4 t_{\varsigma Q}^3} = c^2 t_{\varsigma Q}$$

Verbally interpreted, voltage is a moving distance, a subset of modi.

*The definition of the subset of **voltage** and the $ct$ equation for the subset of voltage, $V_{ltg}$:*

$$V_{ltg} = c^2 t_{\varsigma Q}$$

Conductance is also a subset of modi.

*The definition of the subset of **conductance**, and the $ct$ equation for conductance, $G$:*

$$G = c^2 t_{\varsigma G}$$

Resistance is the inverse of voltage. Resistance is a subset of the physical element of *in*verse *mo*ving *di*stance.

*The definition of the physical element **inmodi**, and the general $ct$ equation for inmodi:*

$$inmodi = \frac{1}{c^2 t}$$

Given the equation $V_{ltg} = IR$ yields an equation for a subset of inmodi, resistance:

$$R = \frac{V_{ltg}}{I} = \frac{c^2 t}{c^4 t^2} = c^{-2} t^{-1}$$

*The definition of the subset **resistance** and the $ct$ equation for resistance, $R$:*

$$R = \frac{1}{c^2 t} = c^{-2} t^{-1}$$

Verbally interpreted, resistance is a numerical value per moving distance, a subset of inmodi.

The following quantum constants are also subsets of inmodi. The following two sections, which consider constant conversions and relationships, details these interconnections further.

$$Z_0' = \frac{1}{c \varepsilon_0} = \frac{1}{c^2 t_f}$$

$$R_K' = \frac{1}{c^2 t_{R_K}}$$

$$G_0^{-1} = \frac{1}{2 c^2 t_{R_K}}$$

Considering the electron charge consumptive volume:

$$t_{V_{Q_e}} = \left(\frac{t_{\varsigma Q_e}^4}{t_{g'}}\right)^{\frac{1}{3}} = \left(\frac{8.4733 \times 10^{-105}}{4.4919 \times 10^{18}}\right)^{\frac{1}{3}} = (1.8863 \times 10^{-123})^{\frac{1}{3}} = 1.2356 \times 10^{-41} \text{s}$$

$$c t_{V_{Q_e}} = 3.7042 \times 10^{-33} \text{m}$$

$$V_{c_{Q_e}} = c^3 t_{V_{Q_e}}^3 = 5.0827 \times 10^{-98} \text{m}^3$$

$V_{c_{Q_e}}$ is the electron charge consumptive volume, the volume of space continuously traveling the speed of light to equal the charge of the electron.

The electron charge mov:



*Eric Mitchell Horn*

$$mov_{Q_e} = cV_{c_{Q_e}} = 1.5238 \times 10^{-89}\,\text{m}^4/\text{s}$$

Checking:

The scat equivalent of electron charge:

$$\varsigma_{Q_e} = mov_{Q_e} * t_{g'} = 6.8447 \times 10^{-71}\,\text{m}^4$$

$$\varsigma_{Q_e} = Q_e * t_{\varsigma_{Q_e}} = 7.1338 \times 10^{-45} * 9.5943 \times 10^{-27} = 6.8444 \times 10^{-71}\,\text{m}^4$$

The following chapter Energy: Part 2 will consider the relationship between electron charge and Planck's constant in detail.

## Energy: Part 2
## Planck Constant Conversion and Relationships
## with Charge and the von Klitzing Constant

*Definition: A **photonic cycle** is an electromagnetic energy cycle, which returns a volume of space to the outer-verse equal to the scat-volume of the charge of the electron.*

For this discussion, and any equation involving Planck's constant, the symbol $\psi$ represents "photonic cycle(s)." The frequency of the photonic cycle is symbolized by the Greek letter $\nu$. The photonic cycle time-length, $t_\psi$ , is the inverse of the photonic cycle frequency. The photonic cycle time-length is usually referred to as the period, symbolized $T$. In this text, the symbol $T$ is reserved for temperature, and the symbol $t$ is reserved for time or time-length and all subsets of the physical element of time will be denoted by $t$ with a right-sided subscript.

*The definition of **photonic cycle time-length**, $t_\psi$ :*

$$t_\psi = \frac{1}{\nu}$$

Conversion of $h$ to $h'$ in units of m⁶/s to equal $c^6 t^5$:

By $ct$ analysis, Planck's constant has units of energy multiplied by time; $c^6 t^4 * t = c^6 t^5$

$$h' = 6.6261 \times 10^{-34} * 2.2262 \times 10^{-19} = 1.4751 \times 10^{-52} \text{ m⁶s⁻¹}\psi\text{⁻¹}$$

$$h' = 1.4751 \times 10^{-52} \text{m⁶s⁻¹}\psi\text{⁻¹} = c^6 t^5 \text{ when } t = 2.8945 \times 10^{-21}\text{s}$$

$$h' = c^6 t_h^5$$

$$t_h = 2.8945 \times 10^{-21}\text{s}$$

$t_h$ is a composite time $= (t_1 \times t_2 \times t_3 \times t_4 \times t_5)^{\frac{1}{5}}$

Each of these $t$'s may also be composite times.

Conversion of $R_K$ to $R_K'$ in units of s/m²:

$$R'_K = \frac{h'}{(e')^2} = \frac{1.4751 \times 10^{-52}}{(7.1338 \times 10^{-45})^2} = 2.8985 \times 10^{36} \text{ s/m}^2 = c^{-2}t^{-1} \text{ when}$$

$$t = 3.8387 \times 10^{-54} \text{s}$$

$$R'_K = \frac{1}{c^2 t_{R_K}}$$

$$t_{R_K} = 3.8387 \times 10^{-54} \text{s}$$

Checking by $ct$ analysis:

$$\frac{c^6 t^5}{(c^4 t^3)^2} = \frac{1}{c^2 t}$$

Considering $E = h\nu$ :

$$h = \frac{E}{\nu} = \frac{Et}{\psi} = \frac{c^6 t^5}{\psi}$$

Notice that $h' * t = c^6 t_h^5 * t = c^6 t^6$

$$c^6 t^6 = (c^3 t^3)^2 = S^2$$

Recall the definition of sess, sustained energy/ space squared:

$$sess = c^6 t^6 = S^2$$

Since $h'$ is a quantum constant, multiplying a quantum constant by a quantum time should yield another quantum constant. The following series of equations will allow development of an additional interpretation of the Planck-Einstein relation $E = h\nu$. This development is important to SPT, since it confirms that energy is a volume of space squared per time squared; i.e. $E = (\frac{V_S}{t})^2$

By observation:

$$h' * t_{R_K} = S^2_{Q_e}$$

$$(1.4751 \times 10^{-52}) * (3.8387 \times 10^{-54}) = 5.6625 \times 10^{-106} \text{m}^6$$

$$S_{Q_e} = 2.3796 \times 10^{-53} \text{m}^3$$

$$S_{Q_e}^2 = 5.6625 \times 10^{-106} \text{m}^6$$

Considering only time factors:

$$t_h^5 t_{R_K} = t_{\varsigma Q_e}^6$$

$$t_{\varsigma Q_e}^3 = \sqrt{t_h^5 t_{R_K}}$$

Continuing:

$$h' * R_K^{-1} = Q_e^2$$

$$c^6 t_{\varsigma Q_e}^6 = S_{Q_e}^2$$

$$S_{Q_e}^2 = h' t_{R_K}$$

$$h' = \frac{S_{Q_e}^2}{t_{R_K} \psi}$$

$$h' * t_{R_K} = \frac{c^6 t_{\varsigma Q_e}^6}{\psi}$$

$$t_{R_K} = \sqrt{\frac{c^6 t_{\varsigma Q_e}^6}{E_P}} = \frac{c^3 t_{\varsigma Q_e}^3}{\sqrt{E_P}} = \frac{c^2 t_{\varsigma Q_e}^3}{A_{\varsigma P}} = \frac{t_{\varsigma Q_e}^3}{t_{\varsigma P}^2}$$

$$t_h^5 = t_{\varsigma Q_e}^3 t_{\varsigma P}^2 = t_{\varsigma P}^4 t_{R_K}$$

$$t_{\varsigma Q_e}^3 = t_{\varsigma P}^2 t_{R_K}$$



Eric Mitchell Horn

$$E = h\nu = \frac{S_{Q_e}^2}{t_{R_K}\psi} * \frac{1\psi}{t_\psi} = \frac{S_{Q_e}^2}{t_{R_K}t_\psi}$$

This confirms that energy is a volume of space squared per time squared.

$$E = \frac{V_S^2}{t^2} = \left(\frac{V_S}{t}\right)^2$$

*The SPT extension of the Planck-Einstein relation:*

$$E = h\nu = \frac{S_{Qe}^2}{t_{R_K}t_\psi}$$

An important difference between this theory and conventional theory is that, for each photonic cycle, a volume of space returns to the outer-verse. Photons are "reverse holes" or hole-complexes in the structure of space through which space particles return from the inner-verse to the outer-verse. Given the equation $E = h\nu$, a higher frequency results in more energy. Since $h$ is a constant and frequency is the only variable in the amount of energy, this is consistent with the proposal that a specific volume of space particles enters the outer-verse for every individual wavelength or photonic cycle of the propagated light wave. It is the number of space particles generated per unit time which accounts for the increase in energy with shorter wavelengths. In effect, huge numbers of space particles are being consumed by an object with mass every second, but energy generates huge numbers of space particles every second as well.

**Hypothesis:** *Photons are hole-complexes propagated as a wave and the media for propagation of the wave is an "ocean" of space particles.*

**Hypothesis:** *Photons are delivery vehicles for the return of a volume of space to the outer-verse.*

By hypothesis and by the above proof:

$$E = \frac{V_S^2}{t^2} = h'\nu = h' * \frac{\psi}{t}, \text{ the value of } h' \text{ must be } \frac{S_{Q_e}^2}{t_{R_K}\psi}$$

$$h' = \frac{S_{Qe}^2}{t_{R_K}\psi} = \frac{2(e')^2}{G_0'\psi} = \frac{2(e')^2}{c^2 t_{G_0}\psi} = \frac{c^6 t_{SQe}^6}{2\alpha t_f\psi} = \frac{2(e')^2}{4\alpha c^2 t_f\psi} = \frac{(e')^2}{2\alpha c^2 t_f\psi} = \frac{(e')^2}{c^2 t_{R_K}\psi} = \frac{R_K'(e')^2}{\psi} = \frac{R_K'c^2 S_{Qe}^2}{\psi}$$

The definition of the **photonic cycle wavelength**, and the $ct$ equation for the photonic cycle wavelength, $\lambda$ :

$$\lambda = ct_\psi$$

The observation that $t_{R_K}$ is the shortest pure time-length thus far obtained in SPT leads to the following hypotheses:

**Hypothesis:** The shortest photonic cycle time, $t_{\psi_0}$ :

$$t_{\psi_0} = t_{R_K}$$

**Hypothesis:** The highest theoretical frequency, $\nu_0$ :

$$\nu_0 = \frac{1\psi}{t_{R_K}} = \frac{1}{3.8387 \times 10^{-54}} = 2.6050 \times 10^{53} \ \psi/s$$

**Hypothesis:** The shortest theoretical wavelength, $\lambda_0$ :

$$\lambda_0 = ct_{R_K} = 1.1508 \times 10^{-45} \text{m}$$

The definition of **Planck energy**, and the $ct$ equation for Planck energy, $E_P$ :

$$E_P = c^6 t_{\varsigma P}^4$$

$$E_P = c^6 t_{\varsigma P}^4 = h'\nu_0 = \frac{S_{Q_e}^2}{t_{R_K}\psi} * \frac{\psi}{t_{R_K}} = \frac{S_{Q_e}^2}{t_{R_K}^2} = 38.426 \ \text{m}^6/\text{s}^2 = 1.7261 \times 10^{20} \text{J}$$

The definition of **Planck moving area**, and the $ct$ equation for Planck moving area, $moar_P$ :

$$moar_P = \sqrt{E_P} = c^3 t_{\varsigma P}^2 = \frac{c^3 t_{\varsigma Q e}^3}{t_{R_K}} = \frac{S_{Q_e}}{t_{R_K}} = 6.1989 \ \text{m}^3/\text{s}$$

The definition of the **Planck scat-equivalent**, and the $ct$ equation for the Planck scat-equivalent, $\varsigma_P$ :



Eric Mitchell Horn

$$\varsigma_P = \frac{c^6 t_{\varsigma P}^4}{c^2} = c^4 t_{\varsigma P}^4$$

$$\varsigma_P = c^4 t_{\varsigma P}^4 = 4.2755 \times 10^{-16} \text{m}^4 = 1920.5 \text{ kg}$$

*The definition of **Planck scat-area,** and the $ct$ equation for Planck scat-area, $A_{\varsigma P}$:*

$$A_{\varsigma P} = c^2 t_{\varsigma P}^2$$

$$A_{\varsigma P} = \sqrt{\varsigma_P} = 2.0677 \times 10^{-8} \text{m}^2$$

Planck scat-area, is the scat-area of the scat equivalent of Planck's constant.

*The definition of **Planck scat-time**, and the $ct$ equation for Planck scat-time, $t_{\varsigma P}$:*

$$t_{\varsigma P} = \left(\frac{E_P}{c^6}\right)^{\frac{1}{4}} = \left(\frac{c^6 t_{\varsigma P}^4}{c^6}\right)^{\frac{1}{4}}$$

$$t_{\varsigma P} = 4.7965 \times 10^{-13} \text{s}$$

Each photonic cycle returns to the outer-verse a volume of space equal to the electron charge scat volume; i.e. the electron charge space, $S_{Q_e}$. $E_P$ is the energy equivalent associated with Planck's constant. Planck energy produces a continuous stream of the electron charge space into the outer-verse.

$$S_{Q_e} = V_{\varsigma Q_e} = c^3 t_{\varsigma Q_e}^3 = c^3 t_{\varsigma P}^2 t_{R_K} = t_{R_K} \sqrt{E_P} = 2.3796 \times 10^{-53} \text{m}^3$$

$$Q_e = c^4 \sqrt{t_h^5 t_{R_K}} = c^4 t_{\varsigma P}^2 t_{R_K}$$

$$t_{\varsigma Q_e}^3 = t_{\varsigma P}^2 t_{R_K}$$

$$Q_e = c^4 t_{\varsigma Q_e}^3 = c(c^2 t_{\varsigma P}^2)(c t_{R_K})$$

The Planck scat-area, $A_{\varsigma P}$ is a moving distance multiplied by a time element. The distance associated with the von Klitzing constant, $c t_{R_K}$, is presumed to be the distance the hole travels at the speed of light in one von Klitzing time-length. This results in a scat-volume equal to $c^3 t^3_{\varsigma Qe}$ in one von Klitzing time-length.

$$c^2 t_{R_K} = R_K^{-1}$$

$$A_{\varsigma P} = Q_e R_K \qquad \text{or}$$

$$Q_e = \frac{A_{\varsigma P}}{R_K} \qquad \text{or summarizing:}$$

$$Q_e = (c^2 t^2_{\varsigma P}) * c * c t_{R_K} = (A_{\varsigma P})(R_K^{-1})$$

This formula reveals the amount of space which replaces the amount of space consumed by the charge of the electron in one electron charge scat-time. Charge is equal to an area of consumption/replacement, moving at the speed of light a very short distance. But, this is a view of charge from the energy or replacement side of the equation. From the discussion in the following sections, this equation can be viewed from a consumptive side also. The charge of the electron is somewhat analogous to the propeller on a ship. The propeller is moving extremely fast relative to the speed of the ship. The charge of the electron is what allows for or causes the electron to move through space. The consumption/replacement of space is balanced, but the direction is not. Utilizing a tornado analogy, charge is a completely formed tornado that hasn't moved through time; i.e. it is a motionless tornado of flowing space particles, a motionless moving volume of space. The air (space) is moving, but the tornado has not moved through time.

Considering the relationship of $E = mc^2$ in the context of an object with mass being an area of consumption associated with the object with mass, $A_\varsigma$ squared, $m_{m^4} = A_\varsigma^2$, and the resultant relationship $E = A_\varsigma^2 c^2$, and considering the units and physical meaning, by hypothesis $E = mc^2$ could be written as:

$$\left(\frac{V_S}{t}\right)^2 = (cA_\varsigma)^2$$

Alternatively:

$$\left(\frac{(ct_\varsigma)^3}{t_\varsigma}\right)^2 = ((ct_\varsigma)^2)^2 c^2 = c^6 t_\varsigma^4$$

Considering $E = h\nu$ :

$$h = \frac{E}{\nu} = \frac{Et}{\psi} = \frac{c^6 t^5}{\psi}$$

Planck's constant is representative of the energy form of the physical element moscop.

$$h * t = \frac{c^6 t^6}{\psi} = \frac{(c^3 t^3)^2}{\psi} = \frac{S^2}{\psi}$$

Recall:

$$ACS_E = Ect$$

$ACS_E$ has a $ct$ expression of $c^7 t^5$ and SI derived units of m$^7$s$^{-2}$. $ACS_E$ reflects the observation that energy travels at the speed of light through time.

$$h * c = \frac{c^7 t^5}{\psi}$$

$$ACS_E = Ect = h\nu ct$$

$$h = \frac{ACS_E}{\nu ct} = \frac{aS^2}{\nu ct} = \frac{S^2}{\nu t^2}$$

$$S^2 = h\nu t^2$$

The definition of the **space equivalent of energy**, $SE_E$:

$$SE_E = t\sqrt{E}$$

$$SE_E = t\sqrt{h\nu} = \sqrt{h\nu t^2} = t\sqrt{E}$$

Checking, using $(\frac{V_S}{t})^2$ as an expression for energy, the relationship to the Planck constant and frequency can be written as:

$$(\frac{V_S}{t})^2 = h\nu$$

$$V_S^2 = h\nu t^2$$

$$V_S = t\sqrt{h\nu} = \sqrt{h\nu t^2} = SE_E$$

Consider this sample Planck constant calculation. Calculating a value for $h'$ in units of m⁶/s:

$$h' = 6.6261 \times 10^{-34} * 2.2262 \times 10^{-19} = 1.4751 \times 10^{-52} \text{ m}^6\text{s}^{-1}\psi^{-1}$$

$$\frac{V_S}{t} = \sqrt{h'\nu}$$

Considering green light with a frequency of $540 \times 10^{12}$ Hz

$$\frac{V_S}{t} = \sqrt{1.4751 \times 10^{-52} * 540 \times 10^{12}} = 2.8223 \times 10^{-19} \text{ m}^3\text{/s}$$

Green light generates $2.822 \times 10^{-19}$ m³/s for each individual photon. The equivalent scat/mass of this single photon of green light:

$$\frac{V_S}{t} = \sqrt{E} = \sqrt{\varsigma c^2}$$

$$V_S = ct\sqrt{\varsigma}$$

The definition of the **space equivalent of scat/mass**, $SE_\varsigma$:

$$SE_\varsigma = ct\sqrt{\varsigma}$$

Continuing:

$$(2.8223 \times 10^{-19})^2 = \varsigma c^2$$

$$\varsigma = 8.8629 \times 10^{-55} \text{ m}^4$$

$$m_{kg} = \frac{m_{m^4}}{2.2262 \times 10^{-19}} = 3.9812 \times 10^{-36} \text{ kg}$$

$3.9812 \times 10^{-36}$ kg consumes the same volume of space from the outer-verse that a single photon of green light returns to the outer-verse.

# Physical Interpretation of Temperature, Pressure and Entropy

**Hypothesis:** *An object with mass can be viewed as energy leaving the outer-verse,* $E_{oo} = c^2 A_\varsigma^2.$

**Hypothesis:** *A localized temperature measurement is the difference between the energy entering the outer-verse,* $E_{io}$, *( energy incoming outer-verse ), versus the energy leaving the outer-verse,* $E_{oo}$, *(energy outgoing outer-verse), per volume of space.*

*The definition of the subset of* **temperature**, *and the* ct *equation for temperature,* $T$:

$$T = c^3 t_\varsigma$$

$$T = \frac{E_{io} - E_{oo}}{(ct)^3} = \frac{E_{net}}{V} \text{ and } T_S = \frac{E_S}{V_S}$$

Simply stated: Space temperature equals space energy per space volume. By $ct$ analysis:

$$T = \frac{c^6 t^4}{c^3 t^3} = c^3 t$$

Temperature will have derived SI units of m³s⁻². Temperature is an accelerated area, a subset of the physical element acar. A temperature measurement is a point measurement of an accelerating volume of space.

At absolute zero, in a given volume of space, the net quantity of space leaving the outer-verse equals [or exceeds] the net quantity of space entering the outer-verse; i.e.:

$$\frac{\Delta V_S}{t} \leq 0$$

$$0 = \frac{E_{io} - E_{oo}}{c^3 t^3}$$

$$(\frac{V_S}{t})^2_{io} \leq (\frac{V_S}{t})^2_{oo} \qquad [(\frac{V_S}{t})^2_{oo} = (cA_\varsigma)^2]$$

*Question:* Is it possible, at any point in space, for the quantity of space being generated by energy to be less than the quantity of space being consumed by the objects with mass at that point in space?
Answer: This is a significant question. Probably not, but possibly within a black hole. This is a question to ponder. The less than (<) symbol may not be applicable, but is placed in the above equation just in case....

Considering the temperature equation, $T_{0K} = c^3 t_{0K} = 0$, temperature is directly proportional to the scat time-length.

$$t_{0K} = 0 \text{ seconds.}$$

The energy loss in a chemical reaction equals the temperature multiplied by the change in entropy:

$$E_{loss} = T\Delta S$$

If temperature has SI derived units of m³/s², and energy has SI derived units of m⁶/s², then entropy must have units of m³ and entropy is a subset of the physical element of space.

*The definition of the subset of **entropy**, and the general equation for entropy, $S$:*

$$S = c^3 t^3$$

The $ct$ equation relating energy, temperature and entropy must be:

$$E = c^6 t^4 = c^3 t * c^3 t^3$$

The $t$ of entropy is a scat-time:

$$E = c^6 t_\varsigma^4 = c^3 t_\varsigma * c^3 t_\varsigma^3$$

An alternative relationship between energy and temperature:

$$E = c^6 t^4 = (c^3 t)^2 * t^2 \text{ or } E = T^2 t^2$$

It is possible to calculate the rate of space increase in the outer-verse per °K:

$$1 \text{ eV} = 1.6022 \times 10^{-19} \text{J} \times 2.2262 \times 10^{-19} \text{m⁴/kg} = 3.5669 \times 10^{-38} \text{ m⁶/s²} = c^6 t^4$$

$$1 eV = 11{,}605 \text{ K}$$

$$c^6 t_E^4 = (c^3 t_T)^2 * t_T^2 = 3.5669 \times 10^{-38} \text{m}^6\text{/s}^2$$

$$t_T = t_E$$

$$t_E = 8.3722 \times 10^{-23} \text{s}$$

$$11{,}605 \text{ K} \times k_T = c^3 t_T$$

$$k_T = \frac{c^3 t_T}{11{,}605} = \frac{c^3 * 8.3722 \times 10^{-23}}{11{,}605} = 0.19438 \text{ m}^3\text{/s}^2 \cdot \text{K}$$

$$1\text{K} = 0.19438 \text{ m}^3\text{/s}^2$$

$$1 \text{ m}^3\text{/s}^2 = 5.1446 \text{ K}$$

The time-length of 1K:

$$c^3 t_{1K} = 0.19438 \text{ m}^3\text{/s}^2 \text{ or}$$

$$t_{1K} = 7.2142 \times 10^{-27} \text{s}$$

Calculating a time-length given a temperature in Kelvin:

$$t_{\varsigma K} = T_K * 7.2142 \times 10^{-27} \text{ (s)} \text{ or } t_{\varsigma K} = \frac{T_K}{1.3862 \times 10^{26}}$$

Calculating the temperature in Kelvin given a time-length:

$$T_K = 1.3862 \times 10^{26} * t_{\varsigma K}$$

A unit of temperature measurement using straight $c^3t$ units:

$$c^3 t = 1 \text{ c}^3\text{t unit of temperature } = 1 \text{ m}^3\text{/s}^2$$
$$t = 3.7114 \times 10^{-26} \text{s}$$

This is equivalent to 5.1446 K. Confirming using the ratio of the two time lengths:

$$\frac{7.2142 \times 10^{-27}}{3.7114 \times 10^{-26}} = 0.19438 \text{ (unit-less)}$$

Checking:

$$t_{1K} = 3.7114 \times 10^{-26} * 0.19438 = 7.2142 \times 10^{-27}\text{s}$$

$$E = T^2 t^2$$

$$T = c^3 t_{11605K} = 11{,}605 \text{ K} \times 0.19438 \text{ m}^3/\text{s}^2 \cdot\text{K} = 2{,}255.8 \text{ m}^3/\text{s}^2$$

$$(2{,}255.8)^2 * (8.3722 \times 10^{-23})^2 = 3.5668 \times 10^{-38}\text{m}^6/\text{s}^2$$

Calculating the volume of space added for 1K at a particular point in space:

$$\frac{V_{S1K}}{t} = c^3 t_{1K}^2 = 1.4023 \times 10^{-27} \text{ m}^3/\text{s}$$

A localized temperature change of 1K means that the difference between the rate of space entering the outer-verse per second and the rate of space leaving the outer-verse per second is 0.19438 m³s⁻²; i.e.:

$$\frac{V_{Si}}{t^2} - \frac{V_{So}}{t^2} = 0.19438 \text{ m}^3\text{s}^{-2} = 1K$$

The cosmic background radiation is about 2.275 K. [7]

$$T = c^3 t_{2.275K} = 2.275 \times 0.19438 = 0.4422 \text{ m}^3/\text{s}^2 = c^3 t$$

$$t_{2.275K} = 1.6412 \times 10^{-26}\text{s}$$

$$T_S = \frac{E_S}{V_S} \qquad \text{(Recall } E_{loss} = T\Delta S; E = ST\text{)}$$

The cosmic energy of space $E_S/\text{m}^3$:

$$E_S = V_S T_S = 1\text{m}^3 \times 0.4422 \text{ m}^3/\text{s}^2 = 0.4422 \text{ m}^6/\text{s}^2$$

$$\frac{0.4422}{2.2262 \times 10^{-19}} = 1.9863 \times 10^{18} \text{ J/m}^3$$

To check the interpretation and units, consider the ideal gas law:

$$PV = nRT$$

Considering the ideal gas constant, $R$:

$$R' = 8.3145 \text{ J/K·mol} \times 2.2262 \times 10^{-19} \text{ m}^4\text{/kg} \div 0.19438 \text{ m}^3\text{/s}^2\text{·K}$$

$$R' = 9.5226 \times 10^{-18} \text{ m}^3\text{/mol}$$

Pressure is force per area and has SI units of N/m², with a $ct$ expression of $\dfrac{c^5 t^3}{c^2 t^2} = c^3 t.$

*The definition of the subset of **pressure** and the equation for pressure, $P$:*

$$Pressure = P = c^3 t_{\varsigma p} \quad \text{(small } p \text{ used to differentiate from Planck scat-time, } t_{\varsigma P})$$

The $ct$ expressions for volume and temperature are $c^3 t^3$ and $c^3 t$ respectively, $n$ is the number of moles and $R$ is the ideal gas constant with SI units of J/K·mol and an expression of $c^3 t^3$/mol. The $ct$ expression analysis for the ideal gas law is:

$$c^3 t * c^3 t^3 = moles * c^3 t^3 \text{ mol}^{-1} * c^3 t$$

Both sides of this equation translate to $c^6 t^4$ which is the $ct$ expression for energy. Temperature and pressure have the same $ct$ expressions. Both result in an accelerating volume of space, and both are subsets of the physical element of acar.

Considering entropy and Boltzmann's constant, $k_B$ :

Entropy is an expression of energy per temperature, $\dfrac{E}{T}$ , with a $ct$ expression of $\dfrac{c^6 t^4}{c^3 t} = c^3 t^3$ or m³. Entropy is therefore a volume of space that is either added to or lost from a system.

$$k_{B'} = 1.3806 \times 10^{-23} \text{ J/K} = 3.0735 \times 10^{-42} \text{ m}^6\text{/s}^2\text{·K} \div 0.19438 \text{ m}^3\text{/s}^2\text{·K}$$

$k_{B'} = 1.5812 \times 10^{-41}$ m³/atom or molecule. This is the same numerical value for entropy which was presented in the Planck constant conversions section.

Checking:

$$k_{B'} = \frac{R'}{N_A} = \frac{9.5226 \times 10^{-18}}{6.0221 \times 10^{23}} = 1.5812 \times 10^{-41} \text{m³/atom or molecule}$$

$k_{B'}$ is the amount of space per atom or molecule required for a gas to have properties of an ideal gas. Referring to the framework "Physical Elements by $c$ and $t$ Analysis," in this application, $k_{B'}$ is the left side of this flowchart from temperature to energy with energy equal to 1 eV.

$$k_{B'} = (ct)^3 = 1.5812 \times 10^{-41} \text{m³}$$

$$1.5812 \times 10^{-41} = c^3 t_{k_B}^3$$

$$t_{k_B} = 8.3722 \times 10^{-23} \text{s}$$

$$t_{1eV} = t_{k_B} = 8.3722 \times 10^{-23} \text{s}$$

$$t_{T_{1eV}} = t_{k_B}$$

Confirming $E = T^2 t^2$:

$$T_{1eV}^2 t^2 = (c^3 t_{k_B})^2 * t_{k_B}^2 = c^6 t_{k_B}^4 = 1\text{eV}$$

$t_{k_B}$ is a scat-time since $E = c^2 \varsigma = c^6 t_{\varsigma k_B}^4$

A published value for kT where T= 298 K is 4.11x 10⁻²¹ J [8]

$$4.11 \times 10^{-21} \text{J} \times 2.2262 \times 10^{-19} \text{m⁴/kg} = 9.15 \times 10^{-40} \text{m⁶/s²}$$

$$k_B' * 298 * 0.19438 = 9.16 \times 10^{-40} \text{m⁶/s²}$$

*The definition of **Planck temperature**, the proposed highest theoretical temperature, and the $ct$ equation for Planck temperature, $T_P$ :*

$$T_P = c^3 t_{\varsigma P} = 1.2924 \times 10^{13} \text{ m}^3/\text{s}^2 \ = 6.6488 \times 10^{13} \text{ K}$$

Considering the electron-volt as a unit of measure for energy:

$$1eV = c^6 t_{k_B}^4 = Q_e * c^2 t = c^4 t_{\varsigma Q_e}^3 * c^2 t$$

Solving for $t$:

$$t = t_{1vlt} = \frac{t_{k_B}^4}{t_{\varsigma Q_e}^3} = \frac{(8.3722 \times 10^{-23})^4}{(9.5943 \times 10^{-27})^3} = 5.5631 \times 10^{-11}\text{s}$$

Checking:

1 joule $= 2.2262 \times 10^{-19}\text{m}^6/\text{s}^2$

1 coulomb = 1 ampere-second $= 4.4525 \times 10^{-26}\text{m}^4/\text{s}$

1 volt = 1 joule per coulomb $= \dfrac{2.2262 \times 10^{-19}}{4.4525 \times 10^{-26}} = 5.0000 \times 10^6 \text{m}^2/\text{s} \ = c^2 t$

$$t_{1volt} = \frac{5 \times 10^6}{c^2} = 5.5631 \times 10^{-11}\text{s}$$

A volt is a manmade unit of moving distance. As such, it is not the smallest moving distance; i.e. not a quantum moving distance.

# Selected Physical Constant Conversions and Relationships

A principle hypothesis of this theory is that all physical elements can be represented mathematically by various combinations of powers of the speed of light and powers of time. A corollary is that all physical constants are related similarly. The question is which constants are most basic and therefore can be used to derive other constants. The constants with the shortest time-lengths are a reasonable starting point. A specific time constant, $t_f$, is fundamental to many of the other quantum constants. Use of the framework "Physical Elements by $c$ & $t$ Analysis" will assist in understanding these relationships.

The notation "$_At_B$" and "$_An_B$" refers to the $t$ and $n$ values between the physical constants on this framework. Many of these conversions are calculated using the basic conversions: $1N = 2.2262 \times 10^{-19}$ m⁵/s² and $1A = 4.4525 \times 10^{-26}$ m⁴/s². See Appendix B; SI Unit to $ct$ Conversions. When converting values given in SI units to a $ct$ expression, the SI value must first be converted to a unit expressed in powers of meters and seconds; i.e. $ct$ compatible units. For the following discussion, the symbol "$\psi$" means "photonic cycle(s)" and $t_\psi$ is the photonic cycle time.

Relationships not presented below may be derived by manipulating these equations to equal one and proceeding to set the resulting equations to equal one another. All physical constants can be expressed in a manner similar to $k_{g''}$ as derived in the section of Gravity: Part 4. Much effort is needed to develop these relationships further.

Many of these relationships have been used in the development of electron charge, and relationships between the quantum constants and charge. Because of the circular nature of these relationships, it is difficult to determine which should be presented first. Some of this information is repeated. It is presented together now with new conversions for convenience and to expand upon in the following chapters.

*The gravity to force constant, $k_{gF}$ :*

$$g''_{Q_e} = k_{g''} * \frac{\varsigma_{Q_e}^2}{c^2 t_d^2} \qquad (\varsigma_Q \text{ is the scat/mass equivalent of charge})$$

$$F''_{Q_e} = k_{e''} * \frac{Q_e^2}{c^2 t_d^2}$$

$$k_{gF} = \frac{g''_{Q_e}}{F''_{Q_e}} = k_{g''} * 4\pi c t_f t_{\varsigma Q_e}^2 = 2.7343 \times 10^{-86} \quad \text{(unit-less)}$$

The time ratio constant of $t_f$ to $t_{g'}$, $k_{t_f g}$ :

$$k_{t_f g} = \frac{t_f}{t_{g'}} = \frac{t_f t_V^3}{t_\varsigma^4} = \frac{2.6301 \times 10^{-52}}{4.4919 \times 10^{18}} = 5.8552 \times 10^{-71} \text{ (unit-less)}$$

The fine structure constant, $\alpha$ :

$$\alpha = \frac{k_{e''} * (e')^2}{h'_b * c} = \frac{1.0093 \times 10^{42} * (7.1338 \times 10^{-45})^2}{7.0385 \times 10^{-45}} = 0.007297 \quad \text{(unit-less)}$$

(0.0072993 accepted)

$$\alpha = \frac{1}{4\pi c t_f} * \frac{(c^4 t_{\varsigma Q_e}^3)^2}{h'_b * c} = \frac{c^6 t_{\varsigma Q_e}^6}{4\pi t_f h'_b} = \frac{S_{Q_e}^2}{2 t_f h'} = \frac{2\pi t_{R_K}}{t_{k_e}} = \frac{\lambda t_{R_K}}{2\varepsilon'_0 t_\psi} = \frac{t_{R_K}}{2 t_f} = \frac{t_{G_0}}{4 t_f} = \frac{t_{\varsigma Q_e}^3}{2 t_{\varsigma P}^2 t_f}$$

$$= \frac{t_{\varphi E_{Q_e}}^2}{2 t_{\varsigma P}^2}$$

$$_\alpha n_c = \frac{1}{\alpha} = \frac{2 t_f}{t_{R_K}}$$

$$t_{R_K} = 2\alpha t_f$$

The constant $\pi$ :

$$\pi = \frac{1}{4 c t_f k_{e''}} = \frac{1}{4\varepsilon'_0 k_{e''}} = \frac{c t_{k_e}}{4\varepsilon'_0} = \frac{t_{k_e}}{4 t_f}$$

$$2\pi = \frac{t_{k_e}}{2 t_f} = \frac{\alpha t_{k_e}}{t_{R_K}}$$

Avogadro number, $N_A$:

$$N_A = (\frac{t_R}{t_{k_B}})^3 \quad \text{(unit-less)}$$

*The tesla unit of measure, $T_u$:*

$$1\,T_u = \frac{c^4 t_{1kg}^4}{s^2 * c^4 t_{1A}^2} = \frac{(7.2456 \times 10^{-14})^4}{s^2 * (2.3478 \times 10^{-30})^2} = 5 \times 10^6 \text{ (unit-less)}$$

*The scat/mass conversion constant, $k_{t_{kgm4}}$:*

$$k_{t_{kgm4}} = \frac{t_{1kg}^4}{t_{1m^4}^4} = \frac{(7.2456 \times 10^{-14})^4}{(3.3356 \times 10^{-9})^4} = 2.2262 \times 10^{-19} \text{ (unit-less)}$$

*The Von Klitzing constant, $R_K'$:*

$$R_K' = \frac{h'}{(e')^2} = \frac{1.4751 \times 10^{-52}}{(7.1338 \times 10^{-45})^2} = 2.8985 \times 10^{36} \text{ s/m}^2 = c^{-2}t^{-1} \text{ when}$$

$$t_{R_K} = 3.8387 \times 10^{-54}\text{s}$$

$$R_K' = \frac{1}{c^2 t_{R_K}} = \frac{1}{2\alpha c^2 t_f} = \frac{2}{G_0'} = \frac{Z_0'}{2\alpha} = \frac{c\mu_0'}{2\alpha}$$

$$t_{R_K} = 2\alpha t_f = \frac{t_{G_0}}{2} = \frac{t_{\varsigma Qe}^3}{t_{\varsigma P}^2}$$

$$R_K' n_{k_{e''}} = \frac{\alpha}{2\pi} = \frac{t_{R_K}}{4\pi t_f}$$

Checking:

$$t_{R_K} = \frac{S_{Qe}^2}{h'} = \frac{5.6625 \times 10^{-106}}{1.4751 \times 10^{-52}} = 3.8387 \times 10^{-54}\text{s}$$

$$t_{R_K} = \frac{t_{\varsigma_{Qe}}^6}{t_h^5} = \frac{7.7997 \times 10^{-157}}{2.0317 \times 10^{-103}} = 3.8389 \times 10^{-54}\text{s}$$

The shortest photonic cycle time-length, $t_{\psi_0}$:

$$t_{\psi_0} = t_{R_K} = 3.8387 \times 10^{-54}\text{s}$$

The fundamental time constant, $t_f$:

$$t_{\psi} = \frac{1\psi}{\nu}$$

$$\nu = \frac{1\psi}{t_{\psi}}$$

$$1\psi = \nu t_{\psi}$$

$$c = \frac{\varepsilon_0'}{t_f} = \lambda\nu = \frac{\lambda}{t_{\psi}}$$

$$\lambda = c t_{\psi}$$

$$\varepsilon_0' = \frac{\lambda t_f}{t_{\psi}}$$

$$\varepsilon_0 t_{\psi} = \lambda t_f$$

$$t_f = \frac{\varepsilon_0' t_{\psi}}{\lambda} = \frac{t_{\varsigma P}^2 t_{R_K}}{t_{\varphi_{B_{Qe}}}^2}$$

$$t_f = 2.6301 \times 10^{-52}\text{s}$$

$$c t_{\varepsilon_0'} = t_f$$

Planck time, $t_P$:

$$t_P = \sqrt{\dfrac{t_{\varsigma Qe}^6}{2\pi t_{g''}^3 t_{R_K}}} = \sqrt{\dfrac{t_h^5}{2\pi t_{g''}^3}} = 5.3912 \times 10^{-44}\,\text{s}$$

*Time relationships:*

$$t_{R_K} = 2\alpha t_f$$

$$t_{G_0} = 2t_{R_K}$$

$$t_{k_e} = 4\pi t_f$$

$$t_{k_e} = \dfrac{2\pi t_{R_K}}{\alpha}$$

$$t_{\varphi_0} = t_{k_J}$$

$$t_{\varsigma P} = \sqrt{2}\,t_{k_J}$$

$$t_{\varsigma P}^2 = 2t_{k_J}^2$$

$$t_{\varphi_{B_{Qe}}}^2 = \dfrac{t_{\varsigma P}^2 t_{R_K}}{t_f}$$

$$t_{\varsigma Qe}^3 = t_{\varsigma P}^2 t_{R_K} = t_{\varphi_{B_{Qe}}}^2 t_f$$

$$t_{\varsigma Qe}^4 = t_{V_{Qe}}^3 t_{g'}$$

$$t_h^5 = 2\pi t_{g''}^3 t_P^2$$

$$t_h^5 = t_{\varsigma Qe}^3 t_{\varsigma P}^2$$

$$t_{\varsigma Qe}^6 = t_h^5 t_{R_K}$$

$$t_{\varsigma Q_e}^6 = t_{\varsigma P}^4 t_{R_K}^2$$

$$t_{\varsigma Q_e}^6 = 2\pi t_{g''}^3 t_P^2 t_{R_K}$$

$$1 = \frac{4\pi t_{\varsigma Q_e}^3 t_f}{t_{\varsigma P}^2 t_{k_e} t_{R_K}}$$

*The inverse conductance quantum, $G_0'^{-1}$:*

$$G_0'^{-1} = \frac{R_K'}{2}$$

$$G_0'^{-1} = \frac{1}{G_0'} = \frac{1}{4\alpha c^2 t_f} = \frac{t_f}{2 t_{R_K} c^2 t_f} = \frac{1}{2 c^2 t_{R_K}} = 1.4493 \times 10^{36}\text{s/m}^2 = c^{-2} t^{-1} \text{ when}$$

$$t = 7.6773 \times 10^{-54}\text{s}$$

$$t_{G_0'^{-1}} = 7.6773 \times 10^{-54}\text{s}$$

$$G_0'^{-1} n_{k_{e''}} = \frac{\alpha}{\pi} = \frac{t_{R_K}}{2\pi t_f}$$

*The conductance quantum, $G_0'$ :*

$$G_0' = \frac{2e'^2}{h'} = \frac{2(7.1338 \times 10^{-45})^2}{1.4751 \times 10^{-52}} = 6.9000 \times 10^{-37}\text{m}^2\text{/s}$$

$$6.9000 \times 10^{-37}\text{m}^2\text{/s} = c^2 t \text{ when } t = 7.6773 \times 10^{-54}\text{s}$$

$$t_{G_0} = 7.6773 \times 10^{-54}\text{s}$$

$$G_0' = c^2 t_{G_0} = 4\alpha c^2 t_f = 2c^2 t_{R_K} = \frac{2}{R_K'}$$

$$G_0' = nc\varepsilon_0' = nc^2 t_f \text{ where } n = \frac{t_{G_0}}{t_f} \text{ or } n = \frac{2 t_{R_K}}{t_f}$$

$$\varepsilon_0' n_{G_0'} = \frac{2t_{R_K}}{t_f}$$

$$t_{G_0} = 4\alpha t_f = 2t_{R_K} = \frac{2t_{\varsigma Qe}^6}{t_h^5} = \frac{2t_{\varsigma P}^4 t_{R_K}^2}{t_h^5} = \frac{2t_{\varsigma Qe}^3}{t_{\varsigma P}^2}$$

$$G_0' t_{\varphi_0'} = \frac{t_{\varsigma Qe}^3}{16\alpha^2 t_f^2} = \frac{t_{\varsigma Qe}^3}{4t_{R_K}^2} = \frac{t_{\varsigma Qe}^3}{t_{G_0}^2} = \frac{\sqrt{t_h^5 t_{R_K}}}{t_{G_0}^2} = \frac{t_{\varsigma P}^2 t_{R_K}}{4t_{R_K}^2} = \frac{t_{\varsigma P}^2}{4t_{RK}}$$

The electric constant, $\varepsilon_0'$ :

$$\varepsilon_0 = 8.8542 x 10^{-12} \text{m}^{-3}\text{kg}^{-1}\text{s}^4\text{A}^2$$

$$\varepsilon_0' = \frac{(8.8542 \times 10^{-12})(4.4525 \times 10^{-26})^2}{2.2262 \times 10^{-19}} \text{ m}$$

$\varepsilon_0' = 7.8847 \times 10^{-44}\text{m} = ct$ when $t = 2.6301 \times 10^{-52}\text{s}$
$t_{\varepsilon_0} = t_f = 2.6301 \times 10^{-52}\text{s}$

$$\varepsilon_0' = ct_f = \frac{\lambda t_f}{t_\psi} = \frac{\alpha^{-1} ct_{R_K}}{2}$$

$$\varepsilon_0' n_{G_0'} = 4\alpha = \frac{2t_{R_K}}{t_f} = \frac{t_{G_0}}{t_f}$$

The magnetic constant, $\mu_0'$ :

$$\mu_0 = 1.2566 \times 10^{-6} \text{ N-A}^{-2}$$

$$\mu_0' = \frac{(1.2566 \times 10^{-6})(2.2262 \times 10^{-19})}{(4.4525 \times 10^{-26})^2} = 1.4111 \times 10^{26} \text{ s}^2/\text{m}^3$$

Checking:

$$\mu_0' = \frac{1}{c^2 \varepsilon_0'} = \frac{1}{c^3 t_f} = \frac{1}{c^2 (7.8847 \times 10^{-44})} = 1.4111 \times 10^{26} \text{ s}^2/\text{m}^3$$

$$\mu_0' = 1.4111 \times 10^{26} \text{ m}^{-3}\text{s}^2 = c^{-3} t^{-1} \text{ when } t = 2.6301 \times 10^{-52}\text{s}$$

$$t_{\mu_0} = t_f$$

$$\mu_0' * nc = Z_0'$$

$$\mu_0' n_{Z_0'} = 1$$

$$\mu_0' * nc = R_K'$$

$$\mu_0' n_{R_K'} = \frac{1}{2\alpha} = \frac{t_f}{t_{R_K}}$$

$$\mu_0' * nc = G_0'^{-1}$$

$$\mu_0' n_{G_0'^{-1}} = \frac{1}{4\alpha} = \frac{t_f}{2 t_{R_K}}$$

*The characteristic impedance of a vacuum, $Z_0'$ :*

$$Z_0' = \mu_0' * c = 4.2304 \times 10^{34} \text{ s}/\text{m}^2 = c^{-2} t^{-1} \text{ when } t = 2.6301 \times 10^{-52}\text{s}$$

$$t_{Z_0} = t_f$$

$$Z_0' = 2\alpha R_K'$$

$$Z_0' = \frac{1}{c^2 t_f}$$

$$Z_0' * nc = k_{e''} \text{ when } n = \frac{1}{4\pi}$$

$$Z_0' n_{k_{e''}} = \frac{1}{4\pi}$$

The inverse characteristic impedance of a vacuum, $Z_0'^{-1}$:

$$Z_0'^{-1} = \frac{1}{Z_0'} = c^2 t_f = 2.3639 \times 10^{-35} \, \text{m}^2/\text{s}$$

The inverse magnetic quantum, or magnetic field strength quantum, $H_0$ (hypothesized) :

$$H_0 = \frac{1}{\mu_0'} = c^3 t_f = 7.0868 \times 10^{-27} \, \text{m}^3/\text{s}^2$$

$$t_{H_0} = t_f$$

$$H_0 = c Z_0'^{-1} = c^2 \varepsilon_0' = c^3 t_f$$

Coulomb's constant, $k_{e''}$ :

$$k_e = 8.9876 \times 10^9 \text{ kg·m}^3/\text{C}^2\text{·s}^2 \text{ or kg·m}^3/\text{A}^2\text{·s}^4$$

$$k_{e''} = \frac{8.9876 \times 10^9 * 2.2262 \times 10^{-19}}{(4.4525 \times 10^{-26})^2} = 1.0092 \times 10^{42} \text{ m}^{-1}$$

$$k_{e''} = 1.0092 \times 10^{42} \, \text{m}^{-1} = c^{-1} t^{-1} \text{ when } t = 3.3050 x 10^{-51} \text{s}$$

$$t_{k_e} = 3.3050 \times 10^{-51} \text{s}$$

$$t_{k_e} = 4\pi t_f$$

$$k_{e''} n_{t^{-1}} = 2\pi = \frac{t_{k_e}}{2 t_f}$$

$$t^{-1} t_\alpha = t_{R_K}$$

Checking:

$$k_{e''} = \frac{1}{c t_{k_e}} = \frac{1}{4\pi \varepsilon_0'} = \frac{1}{4\pi c t_f} = \frac{1}{4\pi} c Z_0'$$

$$\varepsilon_0' = \frac{c t_{k_e}}{4\pi} = c t_f$$

 

*The charge constant, $Q_e$ :*

$$e = 1.6022 \times 10^{-19} \text{ C}$$

$$Q_e = (1.6022 \times 10^{-19})(4.4525 \times 10^{-26}) = 7.1338 \times 10^{-45} \text{m}^4/\text{s}$$

$$Q_e = 7.1338 \times 10^{-45} \text{ m}^4/\text{s} = c^4 t_{\varsigma Qe}^3$$

$$t_{\varsigma Qe} = 9.5943 \times 10^{-27} \text{s}$$

$$t_{\varsigma Qe}^3 = \sqrt{t_h^5 t_{R_K}} = t_{\varsigma P}^2 t_{R_K} = \frac{t_{V_{Qe}}^3 t_{g'}}{t_{\varsigma Qe}}$$

$$Q_e = c^4 \sqrt{t_h^5 t_{R_K}} = c^4 \sqrt{2\alpha t_h^5 t_f} = c^4 t_{\varsigma P}^2 t_{R_K} = c^4 t_{\varphi B_{Qe}}^2 t_f$$

*The charge constant squared, $(e')^2$ :*

$$(e')^2 = Q_e^2 = 5.0891 \times 10^{-89} \text{m}^8/\text{s}^2 = c^8 t_{\varsigma Qe}^6$$

$$t_{\varsigma Qe}^6 = t_h^5 t_{R_K} = 2\alpha t_h^5 t_f = t_{\varsigma P}^4 t_{R_K}^2$$

$$Q_e^2 = c^2 S_{Q_e}^2 = c^8 t_{\varsigma Qe}^6 = c^8 t_h^5 t_{R_K} = 2\alpha c^8 t_h^5 t_f = c^8 t_{\varsigma P}^4 t_{R_K}^2$$


*Josephson constant, $k_J'$ :*

*483598x10⁹ Hz/V*
*1 Hz= 1 $\psi$/s*
*1 V= 5 x 10⁶ m²/s*

$$k_J' = \frac{2e'}{h'} = \frac{2(c^4 t_{\varsigma Qe}^3) t_{R_K}}{c^6 t_{\varsigma Qe}^6} = \frac{2 t_{R_K}}{c^2 t_{\varsigma Qe}^3} = \frac{4\alpha t_f}{c^2 t_{\varsigma Qe}^3} = \frac{t_{G_0}}{c^2 t_{\varsigma Qe}^3} = \frac{t_{G_0}}{c^2 t_{\varsigma P}^2 t_{R_K}} = \frac{2}{c^2 t_{\varsigma P}^2}$$

$$k_J' = 9.6722 \times 10^7 \ \psi/\text{m}^2$$

Checking:

$$k_J' = \frac{2e'}{h'} = \frac{2(7.1338 \times 10^{-45})}{1.4751 \times 10^{-52}} = 9.6723 \times 10^7 \ \psi/\text{m}^2 = \ \psi/ \ c^2 t^2 \text{ when}$$

$$t = 3.3917x10^{-13}\text{s}$$

$$k'_J = \frac{1}{c^2 t^2_{k_J}}$$

$$\frac{1}{t_{k_J}} = \sqrt{\frac{4\alpha t_f}{t^3_{\varsigma Qe}}} = \frac{\sqrt{2}}{t_{\varsigma P}}$$

$$t_{k_J} = \frac{t_{\varsigma P}}{\sqrt{2}} = 3.3912 \times 10^{-13}\text{s}$$

$$k'_J = \frac{2}{c^2 t^2_{\varsigma P}}$$

$$k'_J *_{k'_J} t_{Z'_0} = Z'_0$$

$$_{k'_J} t_{Z'_0} = \frac{t^3_{\varsigma Qe}}{4\alpha t^2_f} = \frac{t^3_{\varsigma Qe}}{t_{G_0} t_f} = \frac{\sqrt{t^5_h t_{R_K}}}{t_{G_0} t_f} = \frac{t^2_{\varsigma P} t_{R_K}}{2 t_f t_{R_K}} = \frac{t^2_{\varsigma P}}{2 t_f}$$

$$k'_J *_{k'_J} t_{R'_K} = R'_K$$

$$_{k'_J} t_{R'_K} = \frac{t^3_{\varsigma Qe}}{8\alpha^2 t^2_f} = \frac{t^3_{\varsigma Qe}}{2 t^2_{R_K}} = \frac{t_{\varsigma P}}{t_{G_0}}$$

The magnetic flux quantum, $\varphi'_0$:

$$\varphi'_0 = \frac{h'}{2e'} = \frac{1.4751 \times 10^{-52}}{2(7.1338 \times 10^{-45})} = 1.0339 \times 10^{-8}\text{m}^2/\psi$$

Checking:

$$\varphi'_0 = \frac{c^6 t^6_{\varsigma Qe}}{2c^4 t^3_{\varsigma Qe} t_{R_K}} = \frac{c^2 t^3_{\varsigma Qe}}{2 t_{R_K}} = 1.0339 \times 10^{-8}\text{m}^2/\psi$$

$$\varphi_0' = 1.0339 \times 10^{-8}\,\text{m}^2/\text{cycle} = c^2 t^2 \text{ when } t = 3.3917 \times 10^{-13}\text{s}$$

$$t_{\varphi_0} = 3.3917 \times 10^{-13}\text{s}$$

Continuing:

$$S_{Q_e} = c^3 t_{\varsigma Q e}^3 = V_{\varsigma Q e} = 2.3796 \times 10^{-53}\,\text{m}^3/\psi$$

$$t_{\varsigma Q e}^6 = \frac{E_P t_{R_K}^2}{c^6}$$

$$t_{\varsigma Q e}^3 = \frac{t_{R_K}\sqrt{E_P}}{c^3} = t_{\varsigma P}^2 t_{R_K} = \frac{t_h^5}{t_{\varsigma P}^2}$$

$$t_{\varsigma P} = \sqrt{\frac{t_{\varsigma Q e}^3}{t_{R_K}}} = t_{\varphi_0}\sqrt{2} = 4.7965 \times 10^{-13}\text{s}$$

$$V_{\varsigma Q e} = t_{R_K}\sqrt{E_P} = 2\alpha t_f \sqrt{E_P}$$

$$\sqrt{E_P} = \frac{V_{\varsigma Q e}}{t_{R_K}} = \frac{V_{\varsigma Q e}}{2\alpha t_f} = \frac{c^3 t_{\varsigma Q e}^3}{2\alpha t_f} = 2c\varphi_0'$$

$$E_P = 4c^2 \varphi_0'^2$$

$$\varsigma_P = 4\varphi_0'^2$$

$$A_{\varsigma P} = 2\varphi_0' = \frac{c^2 t_{\varsigma Q e}^3}{t_{R_K}} = c^2 t_{\varsigma P}^2 = 2.0678 \times 10^{-8}\,\text{m}^2$$

$$\varphi_0' t_{c^2 t^3} = 4\alpha t_f = t_{G_0}$$

$$\varphi_0'^n \sqrt{E_{Q e}} = \frac{4\alpha t_f}{t_{\varsigma Q e}} = \frac{t_{G_0}}{t_{\varsigma Q e}}$$

Checking:

$$t_{\varphi_0} = \sqrt{\frac{t_{\varsigma Q_e}^3}{4\alpha t_f}} = 3.3912 \times 10^{-13}\text{s}$$

$$t_{\varphi_0} = t_{k_J}$$

$$\varphi_0' = c^2 t_{\varphi_0}^2 = \frac{c^2 t_{\varsigma P}^2}{2} = \frac{1}{k_{J'}}$$

*The reduced Planck constant, $h_b'$ :*

$$h_b = 1.0546 \times 10^{-34} \text{ kg·m}^2\text{/s}$$

$$h_b' = h_b * 2.2262 \times 10^{-19} = 2.3477 \times 10^{-53} \text{ m}^6\text{/s} = c^6 t^5 \text{ when } t = 2.0048 \times 10^{-21}\text{s}$$

$$h_b' * c = 7.0385 \times 10^{-45} \text{ m}^7\text{/s}^2 = c^7 t^5 \text{ when } t = 2.0048 \times 10^{-21} \text{ s}$$

$$t_{h_b} = 2.0048 \times 10^{-21}\text{s}$$

$$h_b' = \frac{S_{Q_e}^2}{\alpha t_{k_e}} = \frac{2c^6 t_{\varsigma Q_e}^6 t_f}{t_{R_K} t_{k_e}} = \frac{2c^6 t_{\varsigma P}^4 t_{R_K}^2 t_f}{4\pi t_{R_K} t_f} = \frac{c^6 t_{\varsigma P}^4 t_{R_K}}{2\pi} = \frac{h'}{2\pi}$$

*Boltzmann's constant, $k_B'$ :*

$$k_B = 1.3806 \times 10^{-23} \text{ J·K}^{-1} \text{ or } 8.6173 \times 10^{-5}\text{eV·K}^{-1}$$

$$k_B' = \frac{c^6 t^4}{c^3 t} = c^3 t^3 \text{ / atom or molecule}$$

$$k_B' = \frac{(1.3806 \times 10^{-23})(2.2262 \times 10^{-19})}{0.19438} = 1.5812 \times 10^{-41}\text{m}^3 = c^3 t^3 \text{ when}$$
$$t = 8.3722 \times 10^{-23}\text{s}$$

$$t_{k_B} = t_{\varsigma k_B} = 8.3722 \times 10^{-23}\text{s}$$

*The gas constant, $R'$ :*

$R' = 9.5226 \times 10^{-18}$ m³/mol $= c^3 t^3 / mol$ when $t = 7.0702 \times 10^{-15}$s
$t_R = 7.0702 \times 10^{-15}$s


*The electron-volt unit of measure, $eV'$ :*

$1eV = 3.5669 \times 10^{-38}$m⁶/s² $= c^6 t^4$ when $t = 8.3722 \times 10^{-23}$s

$t_{1eV} = 8.3722 \times 10^{-23}$s

$$1eV = c^3 t_{k_B} k_{B'} = (c^3 t_{k_B})^2 t_{k_B}^2 = c^6 t_{k_B}^4 = T_{1eV}^2 t_{k_B}^2$$

$$t_{1eV} = t_{k_B}$$

*The temperature conversion constant, $k_T$ :*

$k_T = 0.19438$ m³/s²·K
$1$ K $= 0.19438$ m³/s² $= c^3 t_{1K}$
$t_{1K} = 7.2142 \times 10^{-27}$s


*The constant relating $k_{g''}$ to $k_{e''}$ and $k_J'$, a subset of inmointisq, $c^{-1}t^{-2}$:*

$$c^{-1}t^{-2} = \frac{\alpha t_f}{\pi c t_{\varsigma Qe}^3} \qquad \text{(based on } Z_0')$$

$$2 = \frac{t_{R_K}}{\alpha t_f}$$

$$c^{-1}t^{-2} = \frac{2\alpha^2 t_f}{\pi c t_{\varsigma Qe}^3} = \frac{t_{R_K}\alpha^2 t_f}{\alpha t_f \pi c t_{\varsigma Qe}^3} = \frac{\alpha t_{R_K}}{\pi c t_{\varsigma Qe}^3} = \frac{t_{G_0}}{4\pi c t_{\varsigma Qe}^3} = \frac{t_f t_{G_0}}{c t_{k_e} t_{\varsigma Qe}^3} = \frac{1}{2\pi c t_{\varsigma P}^2}$$

$$c^{-1}t^{-2}t_{k_e} = \frac{t_{\varsigma Qe}^3}{t_{G_0} t_f} = \frac{\sqrt{t_h^5 t_{R_K}}}{t_{G_0} t_f} = \frac{t_{\varsigma P}^2}{2 t_f}$$

*The electromagnetic force quantum, $F_{0_{Q_e}}$:*

$$F_{0_{Q_e}} = c Q_e = c^5 t_{\varphi_{B_{Q_e}}}^2 t_f = c^5 t_{\varsigma_P}^2 t_{R_K}$$

$$F_{0_{Q_e}} = 2.1387 \times 10^{-34} \, \text{m}^5/\text{s}^2 = 9.6068 \times 10^{-18} \text{N}$$

$$F_{0_{Q_e}} = Q_e * c = c^5 t_{\varsigma_{Q_e}}^3 = c * c^2 t_f * \varphi_{B_{Q_e}} = c^3 t_f \varphi_{B_{Q_e}} = H_0 \varphi_{B_{Q_e}}$$

$$k_{Q_e F} = c$$

$$k_{Q_e F} = k_{e''} * c^2 t_{k_e}$$

*The strong force quantum, $F_{0_{SF}}$:*

$$F_{0_{SF}} = \frac{2 t_f}{t_{R_K}} * c^5 t_{\varphi_{B_{Q_e}}}^2 t_f = \frac{2 c^5 t_{\varphi_{B_{Q_e}}}^2 t_f^2}{t_{R_K}} = 2 c^5 t_{\varsigma_P}^2 t_f$$

$$F_{0_{SF}} = 2.9306 \times 10^{-34} \, \text{m}^5/\text{s}^2 = 1.3164 \times 10^{-15} \text{N}$$



Eric Mitchell Horn

# Framework Summary of Charge Relationships with Other Quantum Constants

The relationship between charge, $Q$; electric flux, $\varphi_E$; and the electric constant, $\varepsilon_0$ provides a key to discerning other relationships:

$$\varphi_E = \frac{Q}{\varepsilon_0} = c^3 t^2$$

This means that the electric flux, $\varphi_E$, is the square root of energy, $\sqrt{E}$; i. e., a volume of space returning to the outer-verse per time, $= \dfrac{V_S}{t}$ or electric flux is a moving area, $c * c^2 t^2$.

Magnetic flux, $\varphi_B$, is a measure of the total magnetic field passing through an area. The SI unit of measure of magnetic flux is the weber. A weber equals a volt·second equals $c^2 t * t = c^2 t^2 = c^2 t_{\varphi_B}^2$. Magnetic flux is itself an area.

$$EMF = c^2 t = \frac{c^2 t_{\varphi_B}^2}{t}$$

The magnetic flux density, **B**; i.e. the magnitude of the magnetic field, with SI units of wb/m² or tesla, is the net number of field lines passing through a surface, and is unit-less, $\dfrac{c^2 t_{\varphi_B}^2}{c^2 t^2}$. The magnetic field strength, **H**, in ampere/meter, is a directionalized accelerated area or accelerating volume of space. $\mathbf{H} = \dfrac{c^4 t^2}{ct} = c^3 t$. An increase in magnetic field strength yields an increase in temperature. When considering this section, it may be helpful to start with the known relationship, $\varphi_E = \dfrac{Q}{\varepsilon_0} = c^3 t^2$. In order to discern and display relationships, and to further develop the theory, the following framework is presented:

$$c^6 t_{\varsigma P}^4 = 38.425 \ \text{m}^6/\text{s}^2$$
$$c^6 t_{\varphi E_{Qe}}^4 = \varphi_{E_{Qe}}^2 = 8.1972 \times 10^{-03} \ \text{m}^6/\text{s}^2$$

$$c^5 t^4$$
$$moscat_{\varphi E_{Qe}} = c^5 t_{\varphi E_{Qe}}^4 = 2.7343 \times 10^{-11} \ \text{m}^5/\text{s}$$

$$F_{Qe} * t = moscat_{\varphi E_{Qe}}$$
$$t = \frac{t_{\varphi B_{Qe}}^2}{t_f} = 1.2776 \times 10^{25} \ \text{s}$$

$$c^5 t^3$$
$$F_{max_{Qe}} = c Q_e = \varphi_{E_{Qe}} c^2 t_f = \varphi_{E_{Qe}} Z_0^{-1} = c^5 t_{\varsigma P}^2 t_{R_K} = c^2 V_{\varsigma_{Qe}} = 2.1402 \times 10^{-36} \ \text{m}^5/\text{s}^2$$

$$c^4 t^3$$
$$Q_e = A_{\varsigma P} R_K^{-1} = c^2 t_{\varsigma P}^2 * c * c t_{R_K} = 7.1338 \times 10^{-45} \ \text{m}^4/\text{s}$$
$$Q_e = \varphi_{E_{Qe}} * c * t_f = c V_{\varsigma_{Qe}}$$
$$Q_e = \sqrt{E_P} * c t_{R_K}$$
$$I_{Qe} * t = Q_e$$
$$t = t_f = 2.6301 \times 10^{-52} \ \text{s}$$

$$c^4 t^2$$
$$acsp_{Qe} = I_{Qe}$$
$$I_{Qe} = \varphi_{E_{Qe}} * c = c^4 t_{\varphi E_{Qe}}^2 = 2.7143 \times 10^7 \ \text{m}^4/\text{s}^2$$
$$I_{Qe} = \frac{c V_{\varsigma_{Qe}}}{t_f}$$

$$c^3 t^3$$
$$V_{\varsigma_{Qe}} = c^3 t_{\varsigma_{Qe}}^3 = \varphi_{B_{Qe}} c t_f = \varphi_{E_{Qe}} t_f = 2.3796 \times 10^{-53} \ \text{m}^3$$

$$c^3 t^2$$
$$\varphi_E$$
$$\varphi_E = \frac{Q}{\varepsilon_0} = c^3 t^2$$
$$\varphi_{E_{Qe}} = \frac{Q_e}{c t_f} = \frac{V_{\varsigma_{Qe}}}{t_f} = \frac{A_{\varsigma P} c^2 t_{R_K}}{c t_f} = \frac{7.1388 \times 10^{-45}}{7.8848 \times 10^{-44}} = 0.090539 \ \text{m}^3/\text{s} = c^3 t^2$$
$$\left[ t_{\varphi E_{Qe}} = t_{\varphi B_{Qe}} = \sqrt{\frac{0.090539}{c^3}} = 5.7968 \times 10^{-14} \ \text{s} \right]$$

$$H_0 * t = \varphi_{E_{Qe}}$$
$$\left[ t = \frac{t_{\varphi B_{Qe}}^2}{t_f} = \frac{t_{\varsigma P}^2 t_{R_K}}{t_f^2} = 1.2776 \times 10^{25} \ \text{s} \right]$$

$$c^3 t$$

$$H_0 = c^3 t_f = 7.0865 \times 10^{-27}\,\text{m}^3/\text{s}^2 \qquad \varphi_0 = \frac{A_{\varsigma P}}{2} = \frac{\overset{c^2 t^2}{2.0677 \times 10^{-8}}}{2} = 1.0339 \times 10^{-8}\,\text{m}^2$$

$$H_0 = \frac{c^2 V_{\varsigma Q_e}}{\varphi_{B Q_e}}$$

$$\left[ t_{\varphi 0}^2 = \frac{\varphi_0}{c^2} = \frac{1.0339 \times 10^{-8}}{c^2} = 1.1504 \times 10^{-25}\,\text{s}^2 \right]$$

$$\varphi_{B Q_e} = c^2 t_{\varphi_{B Q_e}}^2 = 3.0201 \times 10^{-10}\,\text{m}^2$$

$$\left[ t_{\varphi_{B Q_e}}^2 = 3.3603 \times 10^{-27}\,\text{s}^2 \right]$$

$$\left[ t_{\varphi_{B Q_e}} = t_{\varphi_{E Q_e}} = 5.7968 \times 10^{-14}\,\text{s} \right]$$

$$A_{\varsigma P} = 2.0677 \times 10^{-8}\,\text{m}^2$$

$$\varphi_{B Q_e} = 2\alpha A_{\varsigma P} = A_{\varsigma P} * \frac{t_{R_K}}{t_f} = \frac{S_{Q_e}}{c t_f} = 4\alpha \varphi_0$$

$$A_{\varsigma P} = \frac{S_{Q_e}}{c t_{R_K}}$$

$$\varphi_0 = \frac{S_{Q_e}}{2 c t_{R_K}} = \frac{c^3 t_{\varsigma Q_e}^3}{2 c t_{R_K}}$$

$$\varphi_{B Q_e} = Z_0^{-1} * t$$

$$\left[ t = \frac{\varphi_{B Q_e}}{Z_0^{-1}} = \frac{3.0201 \times 10^{-10}}{2.3638 \times 10^{-35}} = \frac{t_{\varphi_{B Q_e}}^2}{t_f} = \frac{t_{\varsigma P}^2 t_{R_K}}{t_f^2} = 1.2776 \times 10^{25}\,\text{s} \right]$$

$$emf_{Q_o} = Z_0^{-1} = c^2 t_f = \frac{\overset{c^2 t}{\varphi_{B Q_e}}}{t} = 2.3638 \times 10^{-35}\,\text{m}^2/\text{s}$$

$$Z_0^{-1} = c^2 t_f = \frac{c V_{\varsigma Q_e}}{\varphi_{B Q_e}}$$

$$c^2$$

$$\varepsilon_0 = c t_f = \frac{\overset{c t}{V_{\varsigma Q_e}}}{\varphi_{B Q_e}}$$

$$c$$

$$\mathbf{E}$$

$$\mathbf{E{=}B} \times nc$$

$$\overset{t}{Q_e = \varphi_{E_{Q_e}} * c * t_f = A_{\varsigma P} R_K^{-1} = c^2 t_{\varsigma P}^2 * c * c t_{R_K}}$$

$$t_f = \frac{A_{\varsigma P} R_K^{-1}}{c \varphi_{E_{Q_e}}} = \frac{c^4 t_{\varsigma P}^2 t_{R_K}}{c^4 t_{\varphi E_{Q_e}}^2}$$

$$t_f = \frac{t_{\varsigma P}^2 t_{R_K}}{t_{\varphi B_{Q_e}}^2} = \frac{t_{\varphi 0}^2 t_{G0}}{t_{\varphi B_{Q_e}}^2}$$

$$\frac{c}{t}$$

$$1$$

$$\mathbf{B} = \frac{Ft}{ct Q_e}$$

$$\sqrt{2\alpha} = 0.12082$$

$$t^{-1}$$

$$c^{-1} t^{-1}$$

$$k_{e''} = \frac{1}{4\pi c t_f} = \frac{t_{\varphi B_{Q_e}}^2}{4\pi c t_{\varsigma P}^2 t_{R_K}}$$

Multiplying by $\dfrac{c^2}{c^2}$:

$$k_{e''} = \frac{c^2 t_{\varphi B_{Q_e}}^2}{4\pi c^3 t_{\varsigma Q_e}^3} = \frac{\varphi B_{Q_e}}{4\pi V_{\varsigma Q_e}}$$

Verbally $V_{\varsigma Q_e}$ is the electron charge scat-volume. $S_{Q_e}$ is the electron charge space. Both notations refer to the same thing; i.e. $V_{\varsigma Q_e} = S_{Q_e} = c^3 t_{\varsigma Q_e}^3$

$$t_{\varsigma Q_e}^3 = t_{\varphi B_{Q_e}}^2 t_f = t_{\varsigma P}^2 t_{R_K}$$

The magnetic field strength quantum constant, $H_0$, is the inverse of the magnetic constant. $H_0$ is a subset of acar, a directionalized accelerated area (accelerating volume of space.)



*Eric Mitchell Horn*

$$H_0 = \frac{1}{\mu_0} = c^3 t_f$$

$$t^2_{\varphi_{B_{Q_e}}} = \frac{t^2_{\varsigma P} t_{R_K}}{t_f}$$

$$\varphi_{E_{Q_e}} = c^3 t^2_{\varphi_{B_{Q_e}}} = \frac{c^3 t^2_{\varsigma P} t_{R_K}}{t_f} = \frac{H_0 t^2_{\varsigma P} t_{R_K}}{t_f^2}$$

$$S_{Q_e} = \frac{H_0 t^2_{\varsigma P} t_{R_K}}{t_f} = c^3 t^2_{\varsigma P} t_{R_K}$$

$$Q_e = \varphi_{E_{Q_e}} * c t_f = S_{Q_e} * c$$

*Quark Charge, $Q_q$ :*

Since the charge on the electron is divisible, a truer reflection of quantum charge may be quark charge, specifically down quark charge.

$$Q_{q_d} = Q_{q_s} = Q_{q_b} = -\frac{1}{3} * 7.1338 \times 10^{-45} = -2.3779 \times 10^{-45} \text{m\textsuperscript{4}/s} = \text{down quark}$$

charge, strange quark charge, and bottom quark charge, (the negative sign is dropped in the following presentation.)

$$Q_{qd} = c^4 t^3_{\varsigma Q_{qd}} = 2.3779 \times 10^{-45} \text{m\textsuperscript{4}/s}$$

$$Q_{qd} = \frac{1}{3} Q_e = \frac{1}{3} A_{\varphi_{B_{Q_e}}} * c^2 t_f = \frac{1}{3} A_{\varsigma P} * c^2 t_{R_K}$$

$$t_{\varsigma Q_{qd}} = \left(\frac{c^4 t^3_{\varsigma Q_{qd}}}{c^4}\right)^{\frac{1}{3}} = 6.6523 \times 10^{-27} \text{s}$$

$$\varsigma_{Q_{qd}} = c^4 t^4_{\varsigma Q_{qd}} = 1.5819 \times 10^{-71} \text{m\textsuperscript{4}}$$

$$mov_{Q_{qd}} = \frac{\varsigma_{Q_{qd}}}{t_{g'}} = 3.5215 \times 10^{-90} \text{m\textsuperscript{4}/s}$$

$$V_{c_{Q_{qd}}} = \frac{mov_{Q_{qd}}}{c} = 1.1747 \times 10^{-98} \text{m}^3$$

$$t_{V_{Q_{qd}}} = 7.5826 \times 10^{-42} \text{s}$$

$$\frac{A_{\varsigma P}}{3} = 6.8923 \times 10^{-9} \text{m}^2$$

$$c^2 t^2_{\varsigma P_{Q_{qd}}} = 6.8923 \times 10^{-9} \text{m}^2$$

$$t_{\varsigma P_{Q_{qd}}} = \frac{t_{\varsigma P}}{\sqrt{3}} = 2.7692 \times 10^{-13} \text{s}$$

$$t_{\varsigma \varphi B Q_{qd}} = \frac{t_{\varsigma \varphi B Q e}}{\sqrt{3}} = \frac{5.7968 \times 10^{-14}}{\sqrt{3}} = 3.3468 \times 10^{-14} \text{s}$$

*Up quark charge:*

$$Q_{q_u} = Q_{q_c} = Q_{q_t} = +\frac{2}{3} * 7.1338 \times 10^{-45} = +4.7559 \times 10^{-45} \text{ m}^4\text{/s} = \text{up quark}$$

charge, charm quark charge, top quark charge.

Comparing time-lengths for up quark charge, $\varphi_0$ , and down quark charge, as related to Planck scat-time:

$$c^2 t^2_{\varsigma P_{Q_{qu}}} = 1.3785 \times 10^{-8} \text{m}^2$$

$$t_{\varsigma P_{Q_{qu}}} = \sqrt{\frac{2}{3}} * t_{\varsigma P} = 3.9163 \times 10^{-13} \text{s}$$

$$t_{\varphi_0} = \frac{t_{\varsigma P}}{\sqrt{2}} = 3.3916 \times 10^{-13} \text{s}$$

$$t_{\varsigma P_{Q_{qd}}} = \frac{t_{\varsigma P}}{\sqrt{3}} = 2.7692 \times 10^{-13} \text{s}$$

# Fermionic Spin

Considering the Standard Model and the intrinsic angular momentum or spin of fermions, $spin_f$:

$$spin_f = \frac{h_b}{2} = \frac{h}{4\pi}.$$

$$h' = \frac{S_{Q_e}^2}{t_{R_K}\psi}$$

$$1\psi = 2\pi$$

$$h'_b = \frac{S_{Q_e}^2}{2\pi t_{R_K}}$$

$$\frac{h'_b}{2} = \frac{S_{Q_e}^2}{4\pi t_{R_K}}$$

$$spin_f = \frac{S_{Q_e}^2}{4\pi t_{R_K}} = 1.1739 \times 10^{-53}\text{m}^6/\text{s} = \frac{S_{Q_e}^2}{2t_{R_K}\psi} = 7.3755 \times 10^{-53}\text{m}^6/\text{s}{\cdot}\psi$$

$$1.1739 \times 10^{-53}\text{m}^6/\text{s} \div 2.2262 \times 10^{-19} = 5.2731 \times 10^{-35}\text{J}{\cdot}\text{s}$$

$$2t_{R_K} = t_{G_0}$$

$$spin_f = \frac{S_{Q_e}^2}{t_{G_0}\psi}$$

The definition of **fermionic spin** and the $ct$ equation for fermionic spin, $spin_f$:

$$spin_f = \frac{c^6 t_{\varsigma_{Q_e}}^6}{4\pi t_{R_K}} = \frac{c^6 t_{\varsigma_{Q_e}}^6}{t_{G_0}\psi}$$

Checking:

$$spin_f = \frac{h}{4\pi} = \frac{h_b}{2} = 5.2729 \times 10^{-35} \text{J} \cdot \text{s}$$

The fundamental time constant, $t_f$, is derived from the magnetic constant, $\mu_0 = \dfrac{1}{c^3 t_f}$.

Question: What does $t_f$ represent and how does $t_f$ relate to fermionic spin? Is it tied to the intrinsic angular momentum, the spin quantum of fermions in general?

Answer: A previous framework indicates that it is a factor in the charge of the electron. It also appears in Coulomb's constant, $k_{e''} = \dfrac{1}{4\pi c t_f}$. Noting that $\dfrac{1}{4\pi c t_f} = \dfrac{1}{2 * 2\pi c t_f}$, $c t_f$ could be interpreted as the radius of a quantum circle; i.e. one "spin," the spin of the charge of the electron.

Possible hypothesis # 1:

$\varepsilon_0'$ is the radius of the quantum spin of the charge of the electron.

$$r_0 = \varepsilon_0' = c t_f = 7.8848 \times 10^{-44} \text{m}$$

Considering physical meaning with $C_{c0}$ representing the circumference of the smallest circle and $r_{c0}$ representing the radius of the smallest circle, and $d_{c0}$ representing the diameter of the smallest circle:

$$C_{c0} = \frac{c t_{k_e}}{2} = 2\pi c t_f = 2\pi \varepsilon_0$$

$$r_{c0} = \varepsilon_0 = c t_f$$

$$d_{c0} = 2 c t_f = 2\varepsilon_0$$

This assumes $\varepsilon_0$ represents the radius of the smallest circle.

$$F_{Q''} = k_{e''} * \frac{Q_1 Q_2}{c^2 t_d^2} = \frac{Q_1 Q_2}{4\pi c t_f (c^2 t_d^2)}$$

This implies that the space particle cannot be a cube since $2\pi c t_f$ represents the circumference of a circle, and $4\pi c t_f$ may represent the circumference of two circles. If $d_{c_0} = 2c t_f = 2\varepsilon_0$, then:

$$\alpha = \frac{t_{R_K}}{2t_f} = \frac{c t_{R_K}}{d_{c_0}}$$

$\alpha$ may represent the ratio of a distance $(c t_{R_K})$ to the diameter of the smallest circle, $2c t_f$
.

Reconsidering the equation for the force acting between two charges:

$$F_{Q''} = k_{e''} * \frac{Q_1 Q_2}{c^2 t_d^2} = \frac{Q_1 Q_2}{4\pi c t_f (c^2 t_d^2)}$$

Does $2\pi r = c t_f$? Is $c t_f$ the diameter of a quantum circle with $4\pi r$ equal to the diameter of 2 quantum circles? Are there two quantum circles, one in each of two outer-verses? See below.

Possible hypothesis #2:

By the Standard Model, all fermions are spin 1/2 "particles." Could this mean that we actually live in an outer-verse consisting six physical dimensions, two identical outer-verses each with three physical dimensions, but we are able to physically experience only one of the outer-verses at any instantaneous point in time? Could $t_f$ actually represent the time to complete one spin? If so, does $2t_f$ represent the time it takes for completion of a full cycle; that is, starting in one outer-verse, transforming to a second outer-verse, and then back again?

Possible hypothesis #3:

$$\text{Coulomb's constant} = \frac{1}{4\pi c t_f} = 1 \text{ spin per } 2\psi \text{ per } c t_f = 1 \text{ spin per } 2c t_f \text{ per } \psi.$$

i.e. 1 spin per $2\psi$ per $c t_f$ (distance)

Possible hypothesis #4:

From the previous discussion of Coulomb's law, the $4\pi$ term relates a cubic volume to a spherical volume:

$$V = c^3 t^3 = \frac{4\pi[(3ct_f c^2 t_d^2)^{\frac{1}{3}}]^3}{3} = 4\pi c^3 t_f t_d^2 = 4\pi c t_f c^2 t_d^2$$

Possible hypothesis #5:

All or some combination of the above are correct. On initial inspection, some of these hypotheses do not appear to be mutually exclusive. How about it, professionals, what is the correct interpretation?

According to the current development of SPT, the correct hypothesis at this point in time is only:

**Hypothesis:**  $t_f$ and $t_{R_K}$ *have physical meaning.*

# Gravity Part 4: Relationship to Charge, Temperature, $k_{g'}$, and $k_{g''}$

From previous derivation, consider the relationship of charge to $k_g'$:

Recall:

$$k_{g'} = \frac{ct_V^3}{t_\varsigma^4}$$

Checking:

$$t_{V_{Qe}} = 1.2356 \times 10^{-41}\text{s}$$
$$t_{\varsigma_{Qe}} = 9.5943 \times 10^{-27}\text{s}$$
$$k_g' = \frac{c*(1.2356 \times 10^{-41})^3}{(9.5943 \times 10^{-27})^4} = 6.6742 \times 10^{-11}\text{m/s}^2$$

Continuing:

$$e' = Q_e = c^4 t_{\varsigma_{Qe}}^3$$
$$k_{g'}t_{\varsigma_{Qe}} = \frac{ct_{V_{Qe}}^3}{t_{\varsigma_{Qe}}^3}$$

$$t_{\varsigma_{Qe}}^3 = \frac{Q_e}{c^4} = \frac{ct_{V_{Qe}}^3}{k_g' t_{\varsigma_{Qe}}}$$

$$Q_e = \frac{c^5 t_{V_{Qe}}^3}{k_{g'}t_{\varsigma_{Qe}}} = \frac{g' Q_e}{k_{g'}t_{\varsigma_{Qe}}}$$

Checking:

$$k_{g'} = \frac{c^5 t_{V_{Qe}}^3}{Q_e t_{\varsigma_{Qe}}} = \frac{c^5 t_{V_{Qe}}^3}{\varsigma_{Qe}} = \frac{g' Q_e}{\varsigma_{Q_e}} = \frac{ct_{V_{Qe}}^3}{t_{\varsigma_{Qe}}^4}$$

$$c^2 V_c = k_{g'}\varsigma$$

$F''_{Q_e}$ equals the repulsive force acting between two electron charges one meter apart.

$F''_{Q_e}$ will be a factor used to determine the gravitational force constant acting between two scat/masses.

*The definition of $F''_{Q_e}$, and equation and numerical value for $F''_{Q_e}$:*

$$F''_{Q_e} = k_{e''} * \frac{Q_e^2}{1^2} = \frac{c^2 S_{Q_e}^2}{4\pi c t_f} = 1.0093 \times 10^{42} * (7.1338 \times 10^{-45})^2$$

$$= 5.1364 \times 10^{-47} \text{m}^5/\text{s}^2$$

$g''_{Q_e}$ is the gravitational force between the scat equivalents of two electron charges one meter apart ($c^2 t_d^2 = 1$ m²).

*The definition of $g''_{Q_e}$ and equation and numerical value for $g''_{Q_e}$:*

$$g''_{Q_e} = k_{g''} \frac{(c^4 t_{\varsigma Q_e}^4)^2}{c^2 t_d^2} = \frac{(2.9979 \times 10^8) * (4.6846 \times 10^{-141})}{1} = 1.4044 \times 10^{-132} \text{m}^5/\text{s}^2$$

$$g''_{Q_e} c^2 t_d^2 = k_{g''} c^8 t_{\varsigma Q_e}^8$$

The double prime notation of $k_{e''}$ is used to specify that this constant relates the force acting between two similar charges. The constant that relates a single charge with force is the speed of light.

To calculate $k_{g''}$, the repulsive force needs to be subtracted from the equation, (dividing by a factor subtracts that factor from an equation.)

$$F''_{Q_e} = k_{e''} \frac{(c^4 t_{\varsigma Q_e}^3)^2}{c^2 t_d^2}$$

$$F''_{Q_e} c^2 t_d^2 = k_{e''} c^8 t_{\varsigma Q_e}^6$$



*Eric Mitchell Horn*

$$g''_{Q_e} c^2 t_d^2 = k_{g''} c^8 t_{\varsigma Q_e}^8$$

$$\frac{F''_{Q_e} c^2 t_d^2}{g''_{Q_e} c^2 t_d^2} = \frac{k_{e''} c^8 t_{\varsigma Q_e}^6}{k_{g''} c^8 t_{\varsigma Q_e}^8} = \frac{k_{e''}}{k_{g''} t_{\varsigma Q_e}^2} \quad \text{(unit-less)}$$

$$k_{g''} = \frac{g''_{Q_e} k_{e''}}{F''_{Q_e} t_{\varsigma Q_e}^2} = \frac{g''_{Q_e}}{F''_{Q_e} 4\pi c t_f t_{\varsigma Q_e}^2} = \frac{g''_{Q_e}}{F''_{Q_e} c t_{k_e} t_{\varsigma Q_e}^2}$$

The expression "$c t_{k_e} t_{\varsigma Q_e}^2$" contains the units of $k_{g''}$, $c^{-1} t^{-3}$. $\dfrac{g''_{Q e}}{F''_{Q e}}$ is a unit-less constant, the gravity to force constant.

Checking:

$$\frac{F''_{Q_e}}{g''_{Q_e}} = \frac{5.1362 \times 10^{-47}}{1.4044 \times 10^{-132}} = 3.6572 \times 10^{85} \quad \text{(unit-less)}$$

$$\frac{k_{e''}}{k_{g''} t_{\varsigma Q_e}^2} = \frac{1.00926 \times 10^{42}}{(2.9979 \times 10^8)(9.5943 \times 10^{-27})^2} = 3.6572 \times 10^{85} \quad \text{(unit-less)}$$

Combining constants:

$$k_{g''} = \frac{1}{c t_{g''}^3}$$

$t_{g''}^3$ has the numerical value of $\dfrac{1}{c^2}$ :

$$t_{g''}^3 = \# \frac{1}{c^2} \ (s^3)$$

$$t_{g''} = 2.2325 \times 10^{-6} s$$

$$k_{g'} = k_{g''} * c^2 t^2 \text{ when } t^2 = \frac{t_{g''}^3}{t_{g'}}$$

$$t = 1.5738 \times 10^{-18} s$$

$$c^2 t^2 = 2.2262 \times 10^{-19} \mathrm{m^2}$$

Referring to the framework summary, $2.2262 \times 10^{-19} \mathrm{m^2}$ is the area which relates $k_{g''}$ to $k_{g'}$.

Performing $ct$ analysis:

$$\frac{F''_{Q_e}}{g''_{Q_e}} = \frac{k_{e''} * 2.2262 \times 10^{-19}}{k_{g'} * t^2_{\varsigma Q_e}} = \frac{c^{-1}t^{-1}c^2 t^2}{ct^{-1}t^2} \quad \text{(unit-less)}$$

The gravitational force acting between the scat equivalents of two electron charges:

$$g''_{Q_e} = k_{g''} * \frac{(c^4 t^4_{\varsigma Q_e})^2}{c^2 t^2_d} = \frac{(c^4 t^4_{\varsigma Q_e})^2}{c^3 t^3_{g''} t^2_d}$$

The repulsive force acting between two electrons:

$$F''_{Q_e} = k_{e''} \frac{(c^4 t^3_{\varsigma Q_e})^2}{c^2 t^2_d} = \frac{(c^4 t^3_{\varsigma Q_e})^2}{4\pi c^3 t_f t^2_d}$$

Checking:

The gravitational constant for a force acting between two objects with mass a distance $ct_d$ apart:

$$k_{g''} = \frac{g''_{Q_e}}{F''_{Q_e} c t_{k_e} t^2_{\varsigma Q_e}} = \frac{1}{c t^3_{g''}} * \frac{(c^4 t^4_{\varsigma Q_e})^2}{c^2 t^2_d} * \frac{c^2 t^2_d}{k_{e''}(c^4 t^3_{\varsigma Q_e})^2 c t_{k_e} t^2_{\varsigma Q_e}}$$

$$= \frac{1}{c t^3_{g''}} = 2.9979 \times 10^8 \mathrm{m^{-1}s^{-2}}$$

$$g'' * c t^3_{g''} = \frac{c^6 t^6_V t^2_{g'}}{t^2_d} \qquad (t^6_V = t^3_{V1} t^3_{V2}) \, ; \, c t^3_{g''} \text{ is a constant } = 3.3356 \times 10^{-9} \mathrm{m \cdot s^2})$$



Eric Mitchell Horn

$$g' = \frac{ct_V^3}{t_\varsigma^4} * c^4 t_V^3 t_{g'} = \frac{c^5 t_V^6 t_{g'}}{t_\varsigma^4}$$

$$cg' t_\varsigma^4 = c^6 t_V^6 t_{g'}$$

$$g'' * ct_{g''}^3 = \frac{cg' t_\varsigma^4 t_{g'}}{t_d^2}$$

$$g'' = \frac{g' t_\varsigma'^4 t_{g'}}{t_d^2 t_{g''}^3} \quad (t_\varsigma'^4 = \sqrt{t_{\varsigma 1}^4 t_{\varsigma 2}^4} \; ; \frac{t_{g'}}{t_{g''}^3} \text{ is constant } = 4.0371 \times 10^{35} \text{s}^{-2}$$

$$= \frac{1}{2.4770 \times 10^{-36}} = \frac{1}{t^2} \qquad t = 1.5738 \times 10^{-18} \text{s}$$

$$k_{g'} = k_{g''} * c^2 t^2 \text{ when } t = 1.5738 \times 10^{-18} \text{s}$$

$$c^2 t^2 = 2.2262 \times 10^{-19} \text{m}^2$$

$$F_{Q_e}'' = k_{e''} \frac{(c^4 t_{\varsigma Q_e}^3)^2}{c^2 t_d^2} = \frac{(c^4 t_{\varsigma Q_e}^3)^2}{4\pi c^3 t_f t_d^2}$$

$$k_{g''} = \frac{g_{Q_e}''}{F_{Q_e}'' ct_{k_e} t_{\varsigma Q_e}^2} = \frac{g' t_{\varsigma Q_e}^4 t_{g'}}{F_{Q_e}'' ct_d^2 t_{g''}^3 t_{k_e} t_{\varsigma Q_e}^2} = \frac{4\pi c^3 t_f t_d^2 g' t_{\varsigma Q_e}^4 t_{g'}}{c^8 t_{\varsigma Q_e}^6 ct_d^2 t_{g''}^3 4\pi t_f t_{\varsigma Q_e}^2} = \frac{1}{ct_{g''}^3}$$

*The definition of the **gravitational constant acting between two scat/mass quantities,** and the equation, and numerical value for $k_{g''}$ :*

$$k_{g''} = \frac{g_{Q_e}''}{F_{Q_e}'' 4\pi ct_f t_{\varsigma Q_e}^2} = 2.9979 \times 10^8 \text{m}^{-1}\text{s}^{-2}$$

Checking, using Plank's formula for Planck time, $t_P$ (see Planck constant section) to develop $k_{g''}$:

$$k_{g''} = \frac{c^5 t_P^2}{h_b'} = \frac{2\pi c^5 t_P^2 t_{R_K}}{S_{Q_e}^2} = \frac{2\pi t_P^2 t_{R_K}}{c t_{\varsigma Q e}^6} = \frac{2\pi t_P^2}{c t_{\varsigma P}^4 t_{R_K}}$$

Rechecking:

$$g_{Q_e}'' = c^5 t'^3 = \frac{2\pi c^5 t_P^2 t_{\varsigma Q e}^2 t_{R_K}}{t_d^2}$$

$$t'^3 = \frac{2\pi t_P^2 t_{\varsigma Q e}^2 t_{R_K}}{t_d^2}$$

$$F_{Q_e}'' = \frac{c^5 t_{\varsigma Q e}^6}{t_{k_e} t_d^2}$$

$$\frac{g_{Q_e}''}{F_{Q_e}''} = \frac{2\pi t_P^2 t_{R_K} t_{k_e}}{t_{\varsigma Q e}^4}$$

$$k_{g''} = \frac{2\pi t_P^2 t_{k_e} t_{R_K}}{c t_{\varsigma Q e}^6 t_{k_e}} = \frac{2\pi t_P^2 t_{R_K}}{c t_{\varsigma Q e}^6}$$

Considering Planck time, $t_P$:

$$k_{g''} = \frac{k_g'}{c^2 t^2} \quad \text{when } t^2 = 1.5738 \times 10^{-18}\text{s}$$

$$c^2 t^2 = \frac{c^2 t_{g''}^3}{t_{g'}} = 2.2262 \times 10^{-19}\text{m}^2$$

$$t_P' = \sqrt{\frac{h_b' k_{g''}}{c^5}} = \sqrt{\frac{c^6 t_{\varsigma Q e}^6 \, c t_{V_{Q e}}^3 \, t_{g'}}{2\pi c^5 t_{R_K} t_{\varsigma Q e}^4 \, c^2 t_{g''}^3}}$$

$$= \sqrt{\frac{t_{V_{Qe}}^3 t_{\varsigma Qe}^2 t_{g'}}{2\pi t_{g''}^3 t_{R_K}}} = \sqrt{\frac{t_{\varsigma Qe}^6}{2\pi t_{g''}^3 t_{R_K}}} = 5.3914 \times 10^{-44}\text{s}$$

$$t_P^2 = \frac{t_{\varsigma Qe}^6}{2\pi t_{g''}^3 t_{R_K}} = \frac{t_h^5}{2\pi t_{g''}^3}$$

$$k_{g''} = \frac{2\pi t_P^2 t_{R_K}}{c t_{\varsigma Qe}^6} = \frac{2\pi t_{V_{Qe}}^3 t_{\varsigma Qe}^2 t_{g'} t_{R_K}}{2\pi c t_{\varsigma Qe}^6 t_{g''}^3 t_{R_K}} = \frac{t_{V_{Qe}}^3 t_{g'}}{c t_{\varsigma Qe}^4 t_{g''}^3} = \frac{1}{c t_{g''}^3} = 2.9979 \times 10^8 \text{m}^{-1}\text{s}^{-3}$$

Observing:

$$g'' * \frac{1}{k_{g''}} = \left(\frac{scat}{distance}\right)^2 = \left(\frac{c^4 t_V^3 t_{g'}}{c t_d}\right)^2 = S^2$$

If $S^2 = S_{Q_e}^2$, what is the time, $t_d$?

$$\frac{c^8 t_{V_{Qe}}^6 t_{g'}^2}{c^2 S_{Q_e}^2} = t_d^2$$

$$\frac{c^8 t_{\varsigma Qe}^8}{c^2 c^6 t_{\varsigma Qe}^6} = t_d^2$$

$$t_d = t_{\varsigma Qe}$$

Generalizing:

$$S_{\varsigma 1 \varsigma 2}^2 = \frac{\varsigma_1 \varsigma_2}{c^2 t_{\varsigma 1} t_{\varsigma 2}} = \frac{g''}{k_{g''}}$$

Gravity and its relationship with temperature:

*The proposed gravitational loss of temperature, $T_{loss_g}$ (hypothesized):*

$$T_{loss_g} = c^3 t_\varsigma$$

*The proposed relationship of observed gravity, $g'_{obs}$, baseline gravity, $g_{base}$, and measured temperature (hypothesized):*

$$g' = g_{base} = c^2 V_c$$

$$g_{obs} = g_{base} * \left(1 - \frac{T}{T_P}\right)$$

$T_P$ equals Planck temperature, defined previously. Gravity's relationship to other quantum constants will be considered next.

# The Relationship of $k'_g$ to Other Selected Quantum Constants

*The relationship of $k_{g'}$ to $c$:*

$$k_{g'} * t_{g'} = c$$

$$k_{g'} t_c = t_{g'}$$

*The relationship of $k_{g'}$ to $G'_0$ :*

$$G'_0 = k_{g'} * ct^2 = \frac{c^2}{t_{g'}} t^2$$

$$t_{g'} = \frac{t_f}{k_{tfg}}$$

$$G'_0 = \frac{c^2 k_{tfg}}{t_f} t^2$$

$$G'_0 = 4\alpha c^2 t_f$$

$$4\alpha c^2 t_f = \frac{c^2 k_{tfg} t^2}{t_f}$$

$$t_{G_0} = 4\alpha t_f$$

$$t^2 = t_{G_0} * \frac{t_f}{k_{tfg}} = t_{G_0} * t_{g'}$$

$$k_{g'} = \frac{G'_0}{c t_{G_0} t_{g'}}$$

Checking:

$$G'_0 = k_{g'} c t_{g'} t_{G_0} = c^2 t_{G_0}$$

*The relationship of $k_{g'}$ to $\mu'_0$ :*

$$\mu_0' = \frac{k_{g'}}{k_{tfg}c^4} = \frac{ct_{V_{Qe}}^3 t_{g'}}{t_{\varsigma_{Qe}}^4 t_f c^4} = \frac{1}{c^3 t_f}$$

$$k_{g'} = k_{tfg}c^4 \mu_0' = \frac{1}{c^3 t_f} * \frac{t_f t_V^3}{t_\varsigma^4} * c^4 = \frac{ct_V^3}{t_\varsigma^4} = \frac{c}{t_{g'}}$$

The relationship of $k_{g'}$ to $\varepsilon_0'$ :

$$\varepsilon_0' = k_{g'} * t^2$$

$$t^2 = \frac{\varepsilon_0'}{k_{g'}} = \frac{ct_f}{c\dfrac{t_V^3}{t_\varsigma^4}} = \frac{t_\varsigma^4 t_f}{t_V^3} = t_g t_f$$

$$\varepsilon_0' = k_{g'} t_g t_f$$

Checking:

$$\varepsilon_0' = \frac{k_{tfg}c^4}{k_{g'} * c^2} = \frac{k_{tfg}c^2}{k_{g'}} = \frac{k_{tfg}c^2 t_\varsigma^4}{ct_V^3} = \frac{k_{tfg}ct_\varsigma^4}{t_V^3} = k_{tfg}ct_{g'} = ct_f$$

$$k_{g'} = \frac{k_{tfg}c^2}{\varepsilon_0'}$$

$$k_{g'} = \frac{\varepsilon_0'}{t_{g'}t_f} = \frac{c}{t_{g'}} = \frac{\varepsilon_0' t_V^3}{t_f t_\varsigma^4} = \frac{ct_V^3}{t_\varsigma^4}$$

The relationship of $k_{g'}$ to $k_{e''}$ :

$$k_{e''} = \frac{1}{4\pi\varepsilon_0'} = \frac{k_{g'}}{4\pi k_{tfg}c^2} = \frac{t_V^3}{4\pi k_{tfg}ct_\varsigma^4} = \frac{1}{4\pi c t_f}$$

$$k_{e''} = \frac{k_{g'}}{4\pi k_{tfg}c^2}$$

$$k_{g'} = 4\pi k_{tfg}c^2 k_{e''}$$



Eric Mitchell Horn

*The relationship of $k'_g$ to $Z'_0$ :*

$$Z'_0 = \frac{k_{g'}}{k_{tfg}c^3}$$

$$k_{g'} = k_{tfg}c^3 Z'_0$$

Checking:

$$Z'_0 = \frac{k_{g'}}{k_{tfg}c^4} * c = \frac{k_{g'}}{k_{tfg}c^3} = \frac{ct_V^3}{k_{tfg}c^3 t_\varsigma^4} = \frac{t_V^3}{k_{tfg}c^2 t_\varsigma^4} = \frac{1}{c^2 t_f}$$

*The relationship of $k_{g'}$ to $R'_K$ :*

$$k_{g'} = nc^3 R'_K$$

$$n = 8.5459 \times 10^{-73}$$

$$n = 2\alpha k_{tfg}$$

$$k_{g'} = 2\alpha k_{tfg}c^3 R'_K$$

Checking:

$$R'_K = \frac{k_{g'}}{2\alpha k_{tfg}c^3} = \frac{ct_V^3}{2\alpha k_{tfg}c^3 t_\varsigma^4} = \frac{t_V^3}{2\alpha k_{tfg}c^2 t_\varsigma^4} = \frac{1}{2\alpha c^2 t_f} = \frac{1}{c^2 t_{R_K}}$$

*The relationship of $k_{g'}$ to $h'$ :*

$$\alpha = \frac{k_{e''}(e')^2}{h'_b c} = \frac{c^8 t_{\varsigma Qe}^6}{4\pi \varepsilon'_0 h'_b c}$$

$$\varepsilon'_0 = \frac{k_{g'} t_{\varsigma Qe}^4 t_f}{t_{V_{Qe}}^3}$$

$$\alpha = \frac{t_{V_{Qe}}^3}{4\pi k_{g'} t_{\varsigma_{Qe}}^4 t_f} * \frac{2\pi c^8 t_{\varsigma_{Qe}}^6}{h'c} = \frac{c^8 t_{V_{Qe}}^3 t_{\varsigma_{Qe}}^2}{2 k_{g'} c t_f h'}$$

$$k_{g'} = \frac{c^8 t_{V_{Qe}}^3 t_{\varsigma_{Qe}}^2}{2\alpha \varepsilon_0' h'} = \frac{c^8 t_{V_{Qe}}^3 t_{\varsigma_{Qe}}^2 2 t_f}{2 t_{R_K} c t_f h'} = \frac{c^7 t_{V_{Qe}}^3 t_{\varsigma_{Qe}}^2}{t_{R_K} h'} = \frac{c^7 t_{V_{Qe}}^3 t_{\varsigma_{Qe}}^2}{c^6 t_{\varsigma_{Qe}}^6} = \frac{c t_{V_{Qe}}^3}{t_{\varsigma_{Qe}}^4} = \frac{c}{t_{g'}}$$

$$h' = \frac{c^8 t_{V_{Qe}}^3 t_{\varsigma_{Qe}}^2}{2\alpha \varepsilon_0' k_{g'}} = \frac{c^8 t_{V_{Qe}}^3 t_{\varsigma_{Qe}}^2 t_{\varsigma_{Qe}}^4}{2\alpha \varepsilon_0' c t_{V_{Qe}}^3} = \frac{c^6 t_{\varsigma_{Qe}}^6}{2\alpha t_f} = \frac{S_{Q_e}^2}{t_{R_K}}$$

The relationship of $k_{g'}$ to $k_J'$ :

$$k_{g'} = \frac{c^3 t_{V_{Qe}}^3 k_{J'}}{4\alpha t_f t_{\varsigma_{Qe}}} = \frac{2 c^3 t_{V_{Qe}}^3 k_{J'} t_f}{4 t_{R_K} t_f t_{\varsigma_{Qe}}} = \frac{c^3 t_{V_{Qe}}^3 k_{J'}}{2 t_{R_K} t_{\varsigma_{Qe}}} = \frac{2 c^3 t_{V_{Qe}}^3 t_{R_K}}{2 c^2 t_{R_K} t_{\varsigma_{Qe}}^4} = \frac{c t_{V_{Qe}}^3}{t_{\varsigma_{Qe}}^4} = \frac{c}{t_{g'}}$$

$$k_J' = \frac{4\alpha k_{g'} t_f t_{\varsigma_{Qe}}}{c^3 t_{V_{Qe}}^3} = \frac{c t_{G_0}}{c^3 t_{\varsigma_{Qe}}^3} = \frac{2}{c^2 t_{\varsigma_P}^2}$$

# The Space and Energy Relationship

For the following discussion, space, energy, and scat/mass refer to the total space, total energy, and total scat/mass in the outer-verse. The following equations are for illustrative purposes only, since the volume of space in the outer-verse is continually increasing.  A true representation requires calculus.

Consider three functions of time, total space $(S)$, total energy $(E)$ , and total mass/scat $(\varsigma)$ of the outer-verse at time $t$, and $S_i$ is the initial volume of space in the outer-verse:

$$S = S_i + \sqrt{\frac{t}{c}(ACS_E - ACS\varsigma)} = S_i + \sqrt{\frac{t}{c}ct(E - \varsigma c^2)}$$

Simplifying and assuming initial space in the outer-verse equals 0, then:

$$S = t\sqrt{E_{io} - E_{oo}} = t\sqrt{E - \varsigma c^2}$$

If the outer-verse started at a point and t(0) with a "big bang", then $ct$ is the radius of the outer-verse where $t$ is the age of the outer-verse, and the original energy from the "big bang" is at the edge of the outer-verse. The following equation relates an amount of space to energy and scat/mass, assuming that the outer-verse is expanding as a sphere.

$$S = \frac{4}{3}\pi(ct)^3 = t\sqrt{E - \varsigma c^2}$$

Solving for the amount of energy in the outer-verse:

$$E = \varsigma c^2 + \frac{16}{9}\pi^2 c^6 t^4$$

$$E = \frac{S^2}{t^2} = \frac{c^6 t^6}{t^2} = c^6 t^4 = c^3 t^3 * c^3 t = ST = T^2 t^2$$

$$E = \varsigma c^2 + \frac{16\pi^2}{9}T^2 t^2$$

The observation that the universe is expanding at a faster rate with the passage of time; i.e. dark energy is explainable by the relationship $E = ST = T^2 t^2$. The temperature of space, or expressed physically, the acceleration of space, is produced by the return of space particles as photons travel through space. The fact that this acceleration is increasing is the result of the conversion of scat/mass to energy via the fusion process within stars. The total net gain of space in the outer-verse results from a gain in the rate of space accretion from the loss of mass, and with temperature, results in the observations which lead to the hypothesis of dark energy. The total energy of the outer-verse equals the energy associated with the mass of the outer-verse plus the amount of space in the outer-verse times the temperature of the outer-verse.

Considering $t_{g'}$ with a value in SI units of 4.4919 x10¹⁸ s; is 142 billion years, by inference, the age of the outer-verse? This  amount of time is approximately 10 times the currently calculated age of the universe. If the James Webb Space Telescope detects galaxies much farther than are being detected with the Hubble telescope, assumptions in the calculations of the age of the universe may need reconsideration. If $t_{g'}$ is the age of the outer-verse, then the size of the outer-verse can be calculated, and the acceleration due to gravity is predicted to decrease with time.

From previous discussion, the space equivalent of energy, symbolized $SE_E$, is the quantity of space that energy returns to the outer-verse from the inner-verse.

$$SE_E = t\sqrt{E}, \quad \frac{SE_E}{t} = \sqrt{E}$$

The space equivalent of scat/mass, $SE_\varsigma$, is the quantity of space consumed by an object with mass; i.e. the quantity of space leaving the outer-verse and returning to the inner-verse:

$$SE_\varsigma = ct\sqrt{\varsigma} = ctA_\varsigma, \quad \frac{SE_\varsigma}{t} = c\sqrt{\varsigma} = cA_\varsigma$$

Prior to an individual fusion reaction, an object with mass consumes a volume of space equal to $SE_\varsigma = ct\sqrt{\varsigma}$. After that same individual fusion reaction, the amount of space retuned to the outer-verse is $SE_E = t\sqrt{E}$. Conceptually, if the net scat/mass is viewed as a net consumptive area or the net area of the hole-complex, when the mass is changed to energy, the individual hole-complexes are disconnected from each other and the surrounding space particle flows, and the net direction of space particle flow is reversed. The hole-complexes are the sources of the energy wave, and these hole-complexes travel with the energy wave.

**Hypothesis:** *The space particle must have a specific size and shape. There can be no space between space particles.*

This requires the space particles to have the ability to tesselate. Considering the physical implications, with a cube being the simplest of the known possible shapes, and equating all volumes to a cubic volume, then:

$$|ACS_E| = |ACS_\varsigma| = c^7 t^5 = (ct)^3 c^4 t^2 = l^3 c^4 t^2$$

where $l$ is the length of a side of a cube. $l^3$ is the quantity of space in meters cubed of a space particle. Recalling the equation for calculating space gained in the outer-verse from the conversion of mass to energy:

$$SE_E = \sqrt{\frac{t}{c}(ACS_E)} = \sqrt{\frac{t}{c}(l^3 c^4 t^2)} = \sqrt{l^3 c^3 t^3} = (lct)^{\frac{3}{2}}$$

Using Planck units for length and time could yield the volume of the space particle.

$SE_E = (lct)^{\frac{3}{2}} = $ (1.616 x 10⁻³⁵m x 2.998 x 10⁸ms⁻¹ x 5.391 x 10⁻⁴⁴s)³/²

$= 4.22$ x 10⁻¹⁰⁵ m³

The above exercise, which agrees with the published Planck volume, demonstrates the validity of the equations for $SE_E$ and $SE_\varsigma$. This calculation only demonstrates agreement between the published Planck volume, and this calculated volume using Planck length and Planck time. Newton's gravitational constant which Planck employs in the Planck time and Planck length equations represents the gravitational constant between areas of consumption, not two masses. Therefore, this value is not the true size of the space particle.

# The Energy, Gravity, Scat, Temperature, Space, and Distance Quantum

*Question:* What is an appropriate measuring tool for assessing rational numerical values for the above quantum constants?

Answer: Since the space quantum ultimately determines the numerical values of these quantum constants, it would appear on initial review that this would be a good starting point. However, the space quantum, $S_0$ , is purely theoretical, possibly undetectable, and does not currently have an experimental path for provability. The ideal assessment tool would be experimentally verifiable and highly accurate. Currently, no such tool exists.  Perhaps the best currently available indicator for the  rational assessment of a particular numerical value for these quantum constants is temperature.

*Question:* Can the *t* in the temperature be a volume-time?

Answer: No. Please consider the following. By definition and derivation, temperature is related to energy:

$$E = T^2 t^2 = (c^3 t_\varsigma)^2 t_\varsigma^2 = c^6 t_\varsigma^4$$

$$T = \sqrt{\frac{E}{t^2}}$$

$$E = \varsigma c^2$$

$$\varsigma = c^4 t_\varsigma^4 = c^4 t_V^3 t_{g'}$$

$$T = \sqrt{\frac{c^6 t_V^3 t_{g'}}{t^2}}$$

$$c^3 t = \sqrt{\frac{c^6 t_V^3 t_{g'}}{t^2}}$$

The *t* of the *ct* equation for temperature is a composite time and cannot be a volume time.

**Hypothesis:** *Temperature is an accelerated scat-area.*

$$T = \frac{c}{t} * c^2 t_\varsigma^2 = \frac{c}{t} * A_\varsigma$$

**Hypothesis:** *The t in the acar equation:* $acar = c^3 t$ *is a scat-time.*

$$acar = c^3 t_\varsigma$$

**Hypothesis:** *A temperature quantum can be used to assess and calculate possible values for $S_0$.*

**Hypothesis:** *The coldest temperature obtained experimentally can not be colder than the calculated theoretical value of one space particle entering the outer-verse.*

*Question:* Can the quantum the electron charge be used to determine higher level quantum values and the size of the space particle?
Answer: Let's try it!

*The scat/mass equivalent of the charge of the electron:*

$$\varsigma Q_e = c^4 t_{\varsigma Q_e}^3 * t_{\varsigma Q e} = c^4 t_{\varsigma Q e}^4$$

$$(7.1338 \times 10^{-45})(9.5943 \times 10^{-27}) = 6.8444 \times 10^{-71} \text{ m}^4$$

$$mov_{Q_e} = \frac{c^4 t_{\varsigma Q e}^4}{t_g}$$

$$\frac{6.8444 \times 10^{-71}}{4.4919 \times 10^{18}} = 1.5237 \times 10^{-89} \text{ m}^4/\text{s}$$

*The electron charge consumptive volume:*

$$V_{c Q_e} = \frac{mov_{Q_e}}{c} = c^3 t_{V Q e}^3$$

$$\frac{1.5237 \times 10^{-89}}{c} = 5.0826 \times 10^{-98} \text{ m}^3$$

The space particle cannot be any larger than the consumptive volume of the charge of the electron, 5.0826 x 10-98 m³.

*Question:* Does the volume of the space particle, $S_0$ , equal $V_{c_{Q_e}}$ ?

Answer: No. Consider $T_0$ , where $T_0$ is the temperature increase or decrease by one space particle entering or exiting the outer-verse. If this is the case:

$$T_0 = c^3 t_{S_0} = c^3 * t_{S_{Q_e}} = 0.25851 \text{m}^3/\text{s}^2 = 1.3299 \text{ K}$$

In this case $T_0$ is significantly above absolute zero, so it cannot be that $V_{c_{Q_e}}$ represents the size of the space particle. Also, the coldest temperature achieved experimentally is 1 x 10^-10^K [13] , a value significantly below 1.3299K.

*Question:* Would a single space particle; i.e. $S_0$ , entering the outer-verse cause an accelerating volume of space in the outer-verse?
Answer: Intuitively, yes.

*Question:* How is a space particle entering the outer-verse described mathematically?
Answer: $S_0 * c$; that is a moving space particle.  A space particle must move out of the hole-complex in order to be "deposited" in the 3D space of the outer-verse. Therefore, a space particle just as it is entering the outer-verse is moving, which automatically relates to a scat-time.

*Question:* If temperature is related to a scat-time, and this rules out the electron charge consumptive volume, as a determinant of space particle size, what are other candidates for determining space particle size?
Answer: Consider the relationship between energy, charge and voltage:

$$E = QV$$

By SPT theory:

$$E = c^6 t^4 = c^4 t^3 * c^2 t$$

The smallest moving distance demonstrated thus far by SPT is $R_K^{-1}$. A quantum moving distance multiplied by the quantum charge of the electron should yield a quantum of energy.

***Working hypothesis option one,*** $O_1$***:*** *The smallest amount of energy possible is the smallest moving distance multiplied by the charge of the electron.*

*Possible energy quantum #1,* $E_{O_1}$*:*

$$E_{O_1} = Q_e * R_K^{-1} = A_{\varsigma P}^2 * (R_K^{-1})^2 = c^6 t_{\varsigma P}^2 t_{R_K}^2$$

$$E_{O_1} = 2.4612 \times 10^{-81} \text{m}^6/\text{s}^2 = 1.1056 \times 10^{-62} \text{J}$$

Expanding upon this energy quantum allows for further development of related theorized quantum constants.

$$moar_{O_1} = c^3 t_{\varsigma P} t_{R_K} = 4.9610 \times 10^{-41} \text{m}^3/\text{s}$$

$$A_{\varsigma_{O_1}} = c^2 t_{\varsigma P} t_{R_K} = 1.6548 \times 10^{-49} \text{m}^2$$

$$\varsigma_{O_1} = A_{\varsigma_{O_1}}^2 = 2.7384 \times 10^{-98} \text{m}^4$$

$$mov_{O_1} = \frac{\varsigma_{O_1}}{t_{g'}} = 6.0964 \times 10^{-117} \text{m}^4/\text{s}$$

$$g_{O_1} = c * mov_{O_1} = 1.8277 \times 10^{-108} \text{m}^5/\text{s}^2$$

$$S_{O_1} = V_{c_{O_1}} = \frac{g_{O_1}}{c^2} = 2.0335 \times 10^{-125} \text{m}^3$$

$$T_{O_1} = c^3 t_{\varsigma_{O_1}} = c^3 * 1.3569 \times 10^{-33} = 3.6560 \times 10^{-8} \text{m}^3/\text{s}^2 = 1.8809 \times 10^{-7} \text{K}$$

*Question:* Does $S_{O_1}$ represent the size of the space particle, $S_0$?

*Answer:* Probably not. This temperature is greater than the experimental temperature claimed by the Helsinki group in the year 2000 of 1 x 10<sup>-10</sup>K. [13] Also, quarks are known to have fractional values for charge, so the true value of $E_{O_1}$, the smallest amount of energy possible, must be less than that of $E_{O_1}$. Working hypothesis #1 must therefore be false.

*Observation:* $t_{R_K}$ is the shortest photonic cycle time and is a time-length associated with the derived Fermi coupling constant. A reasonable inference is that $(c t_{R_K})^3$ is a factor in the determination of the volume of the space particle. Since $t_{R_K}$ represents the shortest pure time-length derived in this presentation, $c t_{R_K}$ may be the shortest distance.  If this is the case, the shape of the space particle must be determined in order to calculate a volume of the space particle.

*Question:* What is the shape of the space particle?

Answer: The Standard Model predicts that the carrier of the gravitational force, the graviton, will be a spin 2 particle. Considering the space particle as the graviton equivalent, this prediction provides information regarding the shape of the space particle. A spin 1 particle is a particle such that a rotation of $2\pi$ radians, or $360^0$, about one axis will return a particle to show the same image as the original. The sphere is a spin $\infty$ particle, since any rotation will show the same image as the original. The cube is a spin 4 particle since a rotation of $\dfrac{\pi}{2}$ radians, or $90^0$, will show the same image as the original; e.g. $\dfrac{2\pi}{\frac{\pi}{2}} = 4$. This eliminates the cube as the shape of the space particle.

A spin 2 particle will, in effect have 2 mirror images on its face. The spin 2 prediction by the Standard Model and the requirement that the space particle must be able to tessellate, or fill space, yields this hypothesis:

**Hypothesis:** *The shape of the space particle is a golden rhombohedron.*

Of course there are other possibilities for the shape of the space particle, but a golden rhombohedron is the simplest spin 2 space filling particle. The golden rhombohedron can occur in an acute form and an oblate form. The two forms have different volume formulas. Considering the shape to be the acute form and assuming the above hypothesis is correct, it is possible to calculate the volume of the space particle. The formula for calculating the volume of an acute golden rhombohedron, $V_{agr}$ :

$$V_{agr} = \frac{1}{5}\sqrt{10 + 2\sqrt{5}}\,a^3$$ where $a =$ the length of the side of an acute golden rhombohedron.

$$S_0 = V_{agr}$$

**Hypothesis:** *The smallest distance which relates to physical phenomena is $a$, the length of a side of an acute golden rhombohedron.*

**Working hypothesis option two, $O_2$:** *The length on the side of the above space filling acute golden rhombohedron is represented by $ct_{R_K}$ .*

$$a_{O_2} = ct_{R_K} = 1.1508 \times 10^{-45}\text{m}$$



*Eric Mitchell Horn*

The space quantum, $S_0$, based on $t_{R_K}$:

Using $a = ct_{R_K}$

$$S_{O_2} = \frac{1}{5}\sqrt{10 + 2\sqrt{5}}(ct_{R_K})^3 = 0.76085 \times 1.5241 \times 10^{-135} = 1.1596 \times 10^{-135} \text{m}^3$$

$$g_{O_2} = c^2 S_{O_2} = c^2 V_c = c^2 * 1.1596 \times 10^{-135} = 1.0422 \times 10^{-118} \text{m}^5/\text{s}^2$$

$$mov_{O_2} = \frac{g_{O_2}}{c} = 3.4764 \times 10^{-127} \text{m}^4/\text{s}$$

$$\varsigma_{O_2} = mov_{O_2} * t_{g'} = 1.5616 \times 10^{-108} \text{m}^4$$

$$E_{O_2} = c^2 \varsigma_{O_2} = 1.4035 \times 10^{-91} \text{m}^6/\text{s}^2$$

$$moar_{E_{O_2}} = \sqrt{E_{O_2}} = 3.7463 \times 10^{-46} \text{m}^3/\text{s}$$

$$A_{\varsigma_{O_2}} = \frac{\sqrt{E_{O_2}}}{c} = 1.2496 \times 10^{-54} \text{m}^2$$

$$t_{\varsigma_{O_2}} = \left(\frac{\varsigma_{O_2}}{c^4}\right)^{\frac{1}{4}} = (1.9332 \times 10^{-142})^{\frac{1}{4}} = 3.7288 \times 10^{-36} \text{s}$$

$$T_{\varsigma_{O_2}} = c^3 t_{\varsigma_{O_2}} = 1.0047 \times 10^{-10} \text{m}^3/\text{s}^2 = 5.1687 \times 10^{-10} \text{K}$$

This value appears reasonable, but is still greater than the minimum claimed by the Helsinki group. Also, since $t_{R_K}$ is derived from $R_K$, and $R_K^{-1}$ is a moving distance, $t_{R_K}$ must be a scat-time.

**Working hypothesis option three, $O_3$:** *The quantum value of $H_0$ can yield a calculated value of $S_0$.*

$$T_0 = H_0 = c^3 t_f = 7.0865 \times 10^{-27} \text{m}^3/\text{s}^2 = 3.6457 \times 10^{-26} \text{K}$$

This $T_0$ is significantly below the coldest temperature obtained by the Helsinki group, but still above absolute zero.

$$E_{O_3} = H_0^2 t_f^2 = 3.4738 \times 10^{-156} \text{m}^6/\text{s}^2$$

$$\varsigma_{O_3} = \frac{E_{O_3}}{c^2} = 3.8651 \times 10^{-173} \text{m}^4$$

$$m o v_{O_3} = \frac{\varsigma_{O_3}}{t_{g'}} = 8.6044 \times 10^{-192} \text{m}^4/\text{s}$$

$$S_{O_3} = \frac{m o v_{O_3}}{c} = 2.8701 \times 10^{-200} \text{m}^3$$

$$g_{O_3} = c^2 S_{O_3} = 2.5795 \times 10^{-183} \text{m}^5/\text{s}^2$$

$$S_{O_3} = \frac{1}{5}\sqrt{10 + 2\sqrt{5}}\, a^3$$

$$a_{O_3} = \left(\frac{5 S_{O_3}}{\sqrt{10 + 2\sqrt{5}}}\right)^{\frac{1}{3}} = (3.7723 \times 10^{-200})^{\frac{1}{3}} = 3.3538 \times 10^{-67}\text{m}$$

$$t_{a_{O_3}} = 1.1187 \times 10^{-75}\text{s}$$

Considering $H_0$ to be quantum acar (directionalized temperature) yields a space particle size of $2.8701 \times 10^{-200}$m³. Since working hypotheses #1 and #2 yielded values for $T_0$ greater than the coldest temperatures obtained experimentally, they cannot yield the smallest quantum values. Therefore, working hypothesis #3 may yield the proposed quantum values of $E_0$, $g_0$, $\varsigma_0$, $S_0$ and the smallest distance, $a$. Still, these calculated values may not represent the true values for higher level quantum constants.

Finally, consider the following:

$$R_K = \frac{1}{c^2 t_{R_K}}$$

$$R_K^{-1} = c^2 t_{R_K}$$

   Eric Mitchell Horn

$R_K^{-1}$ is a moving distance. Therefore, $t_{R_K}$ cannot be a volume-time. Working option four requires $t_{R_K}$ to be a scat-time.

**Working hypothesis option four, $O_4$:** *The quantum value of $R_K$ can yield a calculated value of $S_0$.*

$$R_K = \frac{1}{c^2 t_{R_K}} = 2.8985 \times 10^{36} \text{s/m}^2$$

$$R_K^{-1} = 3.4501 \times 10^{-37} \text{m}^2/\text{s}$$

$$(R_K^{-1})^2 = c^4 t_{R_K}^2 = 1.1903 \times 10^{-73} \text{m}^4/\text{s}^2$$

$$\varsigma_0 = \varsigma_{R_K} = (R_K^{-1})^2 * t_{R_K}^2 = c^4 t_{R_K}^4 = 1.7540 \times 10^{-180} \text{m}^4 = 7.8788 \times 10^{-162} \text{kg}$$

$$E_0 = E_{R_K} = c^2 \varsigma_0 = c^6 t_{R_K}^4 = 1.5764 \times 10^{-163} \text{m}^6/\text{s}^2 = 7.0810 \times 10^{-145} \text{J}$$

$$mov_0 = \frac{\varsigma_{R_K}}{t_g'} = 3.8847 \times 10^{-199} \text{m}^4/\text{s}$$

$$g_0 = c * mov_{R_K} = \frac{c\varsigma_0}{t_g} = c^2 S_0 = 1.1646 \times 10^{-190} \text{m}^5/\text{s}^2 = 5.2313 \times 10^{-172} \text{N}$$

$$V_{c_0} = S_0 = \frac{mov_{R_K}}{c} = 1.2958 \times 10^{-207} \text{m}^3$$

$$T_0 = c^3 t_{R_K} = 1.0343 \times 10^{-28} \text{m}^3/\text{s}^2 = 5.3210 \times 10^{-28} \text{K}$$

$$a_0 = \left(\frac{5 S_0}{\sqrt{10 + 2\sqrt{5}}}\right)^{\frac{1}{3}} = (1.7031 \times 10^{-207})^{\frac{1}{3}} = 1.1942 \times 10^{-69} \text{m}$$

$$t_{a_0} = \frac{a_0}{c} = 3.9834 \times 10^{-78} \text{s}$$

The above proposed quantum constants obtained from working hypothesis option four are presented in the SPT Summary as a point of discussion. The above values are proposed values pending alternative information. However, the method for determining the higher level quantum constants should be valid. The shape of the space particle which determines the values of $a_0$ and $t_0$ is merely presented for discussion purposes. Remember, $t_{a_0}$, a volume time, is not a true time as we experience time.



*Eric Mitchell Horn*

# Discussion and Conclusion

Humans experience reality in three physical dimensions. The concept of rest mass which is described by the scat equation is shown to have four physical dimensions. SPT demonstrates that space is a quantifiable substance, that matter forms as a result of a consumptive process involving space, that gravity is a direct result of this consumptive process, and that energy is a process that replaces space. For the average person, perhaps the best conception of this reality is to consider this summary:

1.  The "instantaneous" volume of space; i.e. the volume of space continually consumed in the process which forms matter, is $V_c$.
2.  The total area of the consumptive hole-complex is $A_\varsigma$.
3.  The volume of space which is returned per second to equal the volume of space consumed per second in the formation of matter is $\sqrt{E} = cA_\varsigma$.
4.  The concept of time that the average person experiences in everyday life is the time element that describes the space equivalent of matter and energy, $t\sqrt{E}$ and $t(cA_\varsigma)$.
5.  These two subsets of 3D space describe the equivalence between matter and energy in absolute terms. The space consumed through time by matter equals $t(cA_\varsigma)$. The energy through time which is needed to replace this space is $t\sqrt{E}$.
6.  The processes which form matter and result in energy are opposite processes which involve an exchange of a quantity of space between an observable realm and an unobservable realm.


Four observable cornerstones of this theory are:

1.  Energy travels at the speed of light through time.
2.  All matter is attracted to all other matter.
3.  An object with the property of mass can not be accelerated beyond the speed of light.
4.  A moving charge generates a force and the force is directionalized, meaning the electron (or proton) moves in a specific direction in an electric field.

The contemplation of these four cornerstones together allowed development of this Space Particle Theory. Energy traveling at the speed of light through time is a manifestation of accelerated space squared. This fact confirms that space squared exists, which serves to confirm the existence of an inner-verse.

The observational cornerstone that all matter attracts all other matter yields a fundamental hypothesis that an object with the property of mass consumes the media around it. Therefore, that media is a substance, a substance which can be divided, a substance that exists in parts, a substance that exists as a particle; i.e. the space particle or spart.

The third observational cornerstone must indicate the speed at which space particles are traveling when passing from the outer-verse to the inner-verse. This observation directly leads to the fundamental hypothesis that an object with the property of mass must be a volume of space, or space particles traveling at the speed of light through time.

The fourth observational cornerstone allowed the development of the hypothesis of what charge actually is; an area of consumption traveling at the speed of light a very short distance in a particular direction. These four cornerstones were considered before the supporting evidence of Planck units and their conversions. The conversion of Planck units served to confirm the major hypotheses of this theory.

By hypothesis, matter and energy are equal but opposite processes. If the rest-mass of the stars is estimated by observed gravity, then the rest-mass of the outer-verse is highly underestimated since temperature will affect observed gravity. The center of galaxies contain more stars per volume of space and the amount of energy and temperature of that space is greater. This translates to more space returning to the outer-verse near the center of galaxies than in the far reaches of galaxies per volume of space. Inner stars in a galaxy are, in effect, surfing on the volume of space created by the high density of photons in the interior of galaxies. This results in the observed lower orbiting velocities of inner stars versus the higher observed velocities of the outer stars in a galaxy. This provides an explanation for observations that lead to the hypothesis of dark matter.

As time passes, the total amount of scat/mass in the outer-verse decreases as a result of fusion processes within stars. The total amount of energy in the outer-verse increases, resulting in an acceleration in the amount of space in the outer-verse. This provides an explanation for the observation that the outer-verse is extremely large and getting larger at an accelerating rate. The proposed units of temperature, $c^3t$ or m³/s² when combined with the continual conversion of matter to energy indicates an accelerating volume of space, and, by hypothesis, accounts for the observations that lead to the hypothesis of dark energy.

Concerning a system of particle flows, particles entering a flow-stream affect the flow-stream upstream from the entry point. In an accelerating system such as gravity, the effects of space particles entering the outer-verse will be felt "upstream" from the energy source. This theory predicts that for an object with mass parallel with respect to a gravitational field, the bottom of the object with mass will cool more quickly than the top of the object with mass. An experiment can be designed to detect this predicted result.

If a metal bar of sufficient length and mass is hung parallel to earth's gravitational field, and the surrounding temperature is controlled to be exactly the same throughout the length of the bar, heating the bar at the midpoint will result in the temperature of the bottom of the bar rising more slowly than the temperature at the top of the bar. Or, alternatively, if the aforementioned bar is heated such that the temperature is uniform throughout, the bottom of the bar will cool more quickly.

The following observations suggest that a well controlled experiment should be conducted to confirm or deny this hypothesis.

- Crater Lake in Crater Lake National Park in Oregon is a relatively isolated system. The average surface temperature of Crater Lake is 12.8 $^0$C. The temperature of the lake at 300 feet below the surface is about 1$^0$C. There is evidence of areas of warming on the bottom of Crater Lake, due to an underlying magma chamber; i.e. warmer temperatures than in the middle depths of the lake. [9] Since water is a good conductor of heat, what accounts for a colder layer in between two warmer layers of water? This theory provides a possible explanation.

- It is estimated that the average surface temperature of the ocean is about 17 $^0$C. It is estimated that 90% of the ocean volume has a temperature between 0$^0$C and 3$^0$C. [10] The interior core of the planet is estimated to be 6000$^0$C. [12] How can there be an extremely large layer of cold between two warmer layers even though water is a good conductor of heat?

- As a mind experiment, consider a roughly symmetric hill in the Amazon rain forest 300 feet tall, possibly near the city of Manaus, Brazil. Also assume little or no volcanic activity in the geographical area surrounding Manaus. What would be the temperature 300 feet down, level with the base of the hill in the center of the hill? Would it be 27 C which is roughly the average daily temperature of Manaus, ( the average daily temperature varies less than 2 degrees C year round), or would it be less? This theory predicts that the temperature at the bottom of the hole would be significantly less than the average temperature of Manaus. [11]

- A current subject of interest involves the observation that the corona of the sun is much hotter than the surface of the sun. [4] By hypothesis, the rest-mass of the sun consumes energy. The sun produces more energy than it consumes, but this energy is dispersed, or "carried" away at the speed of light. The surface of the sun is expected to be cooler than the corona.

This theory provides an explanation of these observations via this hypothesis:
$$E_{oo} = \varsigma c^2.$$

This theory also provides an alternative explanation for the observations that materials of high density are poor insulators and materials of low density are good insulators, and the observation that most materials expand when heated and contract when cooled.

This relationship may apply to macro sized objects as well. Recent theoretical investigation regarding the internal temperature of the earth suggests a core earth temperature of about 6000 $^0$C. [12] The number of space particles available for consumption decreases with increasing depth. At some depth, the quantity of space particles available is not enough to maintain the stability of the more massive atoms. Nuclear decay results, space particles are generated and become available for consumption in a continuous cycle. It may be that the observed half-life of radioactive isotopes decrease with increasing depth from the earth's surface due to the increasing density of the surrounding matter, and the supply of space particles limited by area, versus the consumption of space particles as a volume. An experiment can be designed to test for this possible phenomenon.

On a subatomic level, each component of an object with mass is a complex of space particle flows. As more neutrons and protons and their constituents are packed in the nucleus, the quantity of space consumed increases proportional to scat/mass. The quantity of space consumed is a volume, but the volume of space available Is limited by surface area. The stability of an atom is thus a function of the ratio of the scat/mass to the surface area. If the ratio of the number of space particles needed for consumption over the number available for consumption is less than some value, instability is the result. In general, more scat/mass per volume translates to less stability.

This theory provides a theoretical foundation for a method of propulsion currently referred to as electromagnetic drive or EM drive. [6] Many non-gravitational forces are fundamentally directionalized gravity. Consider a neutron in free space at absolute zero. The consumptive pattern of that neutron would theoretically be such that the consumption of space in all directions is equally represented. The only method by which motion can be achieved is to change the consumptive pattern of the neutron to consume proportionally more space particles in the direction of motion.

Electromagnetic controls can be used to influence the consumptive pattern of an object with mass. Also, specific chemical bonding structures within a metal or metal-like material or layers of different metals can be used to affect an objects consumptive pattern. The combination of electromagnetic controls and chemical bonding structure of a material can be combined to influence the consumptive pattern of a substance. The resultant consumptive pattern can move an object through space.

The method by which motion is achieved is more correctly viewed as a pull/push (consumption/energy) process, a subatomic version of the traditional action/reaction of Newtonian mechanics. By hypothesis, gravity and the EM drive may have the same physical process for observed motion. The difference is that, for regular matter, the

 *Eric Mitchell Horn*

atoms are consuming space particles from all directions, whereas individual atoms in the metal of an idealized EM drive can be oriented to preferentially consume space particles from a particular direction and that direction is controllable by electromagnetic means.

If a flat surface is considered, one side of the material (metal) could preferentially consume space particles from the desired direction of travel and the other side of the material could serve as an "energy dump" where space particles would preferentially be returned to the outer-verse. It is probable that these complex metals would inherently be high temperature super conductors of two or more different layers of metal. An external electromagnetic field as well as internal fields created by individual atoms could control the orientation of the consumptive pattern of the atoms. A possible design of an EM drive could have flat surfaces. Applying this theory to NASA's EM drive could lead to the development of a more efficient EM drive, one capable of taking mankind to the stars in a human lifetime.

A fundamental hypothesis of this theory is that the consumption of space particles by an object with mass allows for motion. Any change in the motion of an object with mass involves inertia. A change in motion requires a change in the consumptive pattern of the rest-mass. Energy, a return flow of space particles, is required to change the consumptive pattern of a rest-mass, as well as provide a replacement of space on the backside of the directionalized consumptive process. This results in the observed property of inertia.

Every specific calculation in this presentation is directly dependent on an accurate value of Newton's gravitational constant, $G_{kN}$. By hypothesis, since the conversion of mass in kilogram to scat/mass in m⁴ is derived from $k_{g'}$, if $G_{kN}$ is not measured at absolute 0, then the exact value of $k_{g'}$ is subject to considerable error which in turn introduces error into the conversion of kilograms to m⁴. This error is compounded through the process of raising scat-times to various powers. A truly accurate measurement of $k_{g'}$ requires that all mass and energy in the surrounding field be at absolute 0, (an impossibility?).

When an object with mass is viewed as a volume of space traveling at the speed of light through time, how is a mass standardized if time can be any value? This presentation standardizes that time to $t_{g'}$, where $t_{g'}$ is derived from maximum momentum $p_{max}$ or the gravitational constant, $k_{g'}$.

SPT has demonstrated that physical elements and physical constants can be represented by combinations of powers of the speed of light and time. The unit of the kilogram, as a fixed entity for the quantification of matter, is unnecessary and may be an impediment to a truer understanding of the physical nature of the universe. In light of the equations presented, the use of the property of mass to quantify matter should be reviewed since the property of mass as currently defined changes with changing conditions. SPT reveals that the physical element that properly quantifies matter is scat. The mass in Newton's gravity equation is demonstrated to relate to scat-areas associated with matter, not mass per se. The mass in Newton's force equation and Einstein's energy equation relate directly to scat.  The mass in Einstein's relativity equation relates to the property of mass. The conflict between the property of mass and the use of this property to quantify matter has prevented the unification of gravity with the other elementary forces. Given the evidence provided, a narrower definition of mass is warranted. Mass cannot be both a quantifier of matter and a property of matter. Because of this conflict, this presentation has left mass as the property identified in Einstein's relativity equation, and the symbol $m$ as denoting this property.

The equations for temperature/pressure, entropy, charge, current, voltage, resistance, scat/mass and energy give greater understanding to the true physical nature of these physical elements. The physical element of sustained energy space squared, sess, is proven to exist by two means: the alternative form of the Planck-Einstein relation, $S_{Q_e}^2 = E_p t_{R_K}^2$, and the observation that energy traveling at the speed of light through time is accelerated sustained energy space squared, $Ect = c^7 t^5 = c^6 t^6 * \dfrac{c}{t}$.

What is the true shape of the space particle? This shape should be discernible from quantum constant relationships, ascertained by the existence of a time-length, or a time-length ratio, a volume per area, etc. The Standard Model predicts that the particle carrying gravitational force is a spin 2 particle. This would appear to rule out the cube since the cube is a spin 4 particle. Does $a_{O_4}$, derived in the working hypothesis option four from a previous chapter, represent the shortest distance; i.e the length of a side of a hypothesized space particle? A golden rhombohedron is the simplest space filling spin 2 particle. The acute shape was used to calculate $a_0$ and $t_{V_0}$.

However, a rhombic dodecahedron is also a strong candidate for the shape of the space particle. The abundance of $\sqrt{2}^2 = 2$ in various formulas suggest this shape as a strong possibility, since the long diagonal of a face of a rhombic dodecahedron is $\sqrt{2}$ times the short diagonal. The face transitive property of the rhombic dodecahedron allows the space particle to rotate about an axis and still remain in the same relative place in space. Further effort is needed to confirm or deny these possibilities.

Two of the most significant discoveries revealed by SPT are the fundamental time constant, $t_f$, and the time constant associated with the von Klitzing constant, $t_{R_K}$. The discovery of these two time constants allowed for further development of relationships between other quantum constants. What does $t_{R_K}$ represent?  Does $t_{R_K}$ represent a minimum time necessary to replace a volume of space in the outer verse?

What does $t_f$ represent?  Does $ct_f$ represent the radius of a quantum circle?  Given quantum mechanics and the Standard Model, since the electron is classified as a spin 1/2 particle, does $t_f$ reflect the time-length for the charge of the electron to complete two revolutions? Could this reflect a supposition that $2t_f$ is the time to complete two cycles? Since all fermions have quantum spins of 1/2, is it possible that there are in fact multiple outer-verses and that we are in one three dimensional (physical) outer-verse at any instantaneous point in time, transforming from one 3D outer-verse to the next?  In effect, is our existence a movie of a continuous transformational process between 3D outer-verses? It is probable that professionals will be able to answer these questions definitively, possibly in a relatively short period of time.

Subjects for future consideration include the possibility that matter in the outer-verse may be energy in the inner-verse and energy in the outer-verse may be matter in the inner-verse, or matter in the outer-verse may be antimatter in the inner-verse. The continuing determination of numerical values as ratios of quantum constants and other constants and powers of time is an area for future research. A continuing effort to ascertain the ratios that determine the numerical values of the quantum constants would further relate SPT to the Standard Model. Adding calculus to the concepts presented and inclusion of Einstein's relativity which considers the property of mass into a physical element is needed to complete the theory.

SPT does not consider the number of dimensions in the universe, although it could be hypothesized that the number of dimensions equates with the number of $c$'s necessary to describe an observed physical phenomenon. For example, the simple observation that gravity can move a scat/mass a distance yields this hypothesis:

$$? = c^5 t^3 * c^4 t^4 * ct = c^{10} t^8 \text{ and a 10 dimensional universe.}$$

The observation that energy can move a scat/mass a distance yields this hypothesis:

$$? = c^6 t^4 * c^4 t^4 * ct = c^{11} t^9 \text{ and an 11 dimensional universe.}$$

These are unnamed physical elements. What is the minimum number of $c$'s necessary to describe how the universe works? Much research in the pursuit of this knowledge is needed to complete the theory.  Do we, in fact, live in an eleven dimensional universe composed of three physical sub-verses, each sub-verse having three physical dimensions which share two time dimensions? Do we transform between the three sub-verses so quickly (in a time-length equal to $t_f$ or $t_{R_K}$) that we are unaware of our transformation? At any instantaneous moment in time, are we entirely physically present in one sub-verse or is it possible that 1/3 of each of us exists in each of the sub-verses?

While the calculated value of quantum gravity may not be correct because of incomplete information, a true value of SPT lies in the gravitational equation $g' = c^2 V_c$ which provides a direct link to quantum gravity.

Are the physical constants constant? This is another question which needs serious consideration. SPT does not consider this question. If $t_{g'}$ represents the age of the universe, then $k_{g'}$ will not be constant with time.

Careful consideration should be given to the future development of symbolism and terminology. The symbolism and terminology presented in SPT should be reviewed and changed as appropriate. With today's technology, perhaps one symbol incorporating the symbol for space, $S$, the symbol for the speed of light, $c$, and the symbol for time, $t$, could be designed and used in equations in place of the symbol $m$ or $\varsigma$ to represent the physical element of scat/mass as defined.

When considered in its entirety, SPT in some ways may represent a paradigm shift when thinking about the processes governing the universe. However, there are in fact only a few necessary revisions to current theories. A modification to Newton's gravitational equation enables proper adjustments to Planck's quantum mechanics. Also, it is necessary to comprehend that Einstein's Theory of Relativity considers the property of mass, not the quantity of rest mass; and that the property of mass and the quantity of rest mass are two different entities entirely. Likewise, the view that mass and energy are the same is misguided. This is akin to believing that space and gravity are the same. It would be more appropriate to state that temperature, pressure, and magnetic field strength are the same. This is the equivalent of the recognition of the various forms of energy; i.e. heat, electrical, mechanical, electromagnetic, nuclear, etc. A change in the definition of units, eliminating the unit of the kilogram, in favor of $ct$ compatible units is recommended for development of a clear understanding of how the universe works.

Finally, the foundation of SPT relies on the author's firm commitment to this fundamental hypothesis which shall end this discussion.

***Physical phenomena have physical explanations.***



*Eric Mitchell Horn*

# Author's Note

A significant portion of this theory has been developed during the early morning hours from 2:00 a.m. to 5:00 a.m. I find these quiet morning hours of uninterrupted thought most conducive to the creative process. Within this presentation there are errors in thought and in text. This text has been entirely created on an iPad. While the creative process at this time of day is best for me, the mechanical processes of typing on an iPad and performing calculations on a calculator are not. I have tried to identify and correct all errors. However, a few may remain. I will be indebted to readers who discover and report these errors to me.

As a nonprofessional, I would have preferred working with a professional to develop this Space Particle Theory. I was unsuccessful in my attempts to engage the few academic professionals who I approached. Although the probability of a nonprofessional developing a theory of any importance to the general field of physics is negligible, it cannot be zero. I believe that SPT is a significant development in the understanding of the workings of the universe, and can be used by professionals to further this understanding.

Since this text is a product of independent thought, not of experimental research or an extensive reading of the literature, I may have unintentionally failed to cite relevant sources. All information which I gathered to develop this theory; e.g. the mass and charge of the electron, the values for the known quantum constants, etc. are widely available on the internet, which was my only true source of information. Therefore, a comprehensive list of bibliographic references has not been provided. The hypotheses and equations presented have not been reviewed by professional theoretical physicists. The reader is encouraged to think about the theory and perform the calculations independently.

# Acknowledgements

I wish to thank my wife Terri for her unwavering support. Many wives would have the view that the development of an all encompassing theory of how the universe works by their spouse was a useless endeavor. Brian Price was the first to show any real interest in the theory. I thank him for his suggestion. I also wish to thank Matt Beekman, Ph.D. Assistant Professor at Cal Poly (formerly at Oregon Institute of Technology) and Sean Sloan, Ph.D. Associate Professor at Oregon Institute of Technology for their patience listening to early conceptual formulations of the theory. I am extremely grateful for their generosity in sharing their time with me. Ed Silling, Ph.D. has been extremely helpful in many ways. Richard Anderson helped fashion an early temperature experiment which provided the impetus to develop the theory mathematically. Charles Schink has proved invaluable in helping me navigate the unfamiliar realm of technology and its utility in sharing this theory with a wider audience. I thank Thomas Burns, Ph.D. for his editing expertise which has proven invaluable. This book would not be the same without his input. I am grateful to the individual members of the Klamath Writer's Group: William Huntsman, Jean Lamb, Shirley Leggett, and Janeane Kansaku for their suggestions. I express my deep gratitude to these friends for their support. I also wish to thank Patricia Marshall, Kim Harper-Kennedy, Melissa Thomas, and Nina Leis at Luminare Press for their help in making this book a reality.

# Appendix A
# Selected Time-lengths

| Time lengths: | $ct$ expression |
|---|---|
| $t_{g'} = 4.4919 \times 10^{18}\,\text{s}$ | $t$ |
| $t_{g''} = 2.2325 \times 10^{-6}\,\text{s}$ | $c^{-1}t^{-3}$ |
| $t_{1mtr} = 3.3356 \times 10^{-9}\,\text{s}$ | $ct$ |
| $t_{1Vlt} = 5.5632 \times 10^{-11}\,\text{s}$ | $c^2 t$ |
| $t_{\varsigma P} = 4.7965 \times 10^{-13}\,\text{s}$ | $c^6 t^4$ |
| $t_{\varsigma P_{Qqu}} = 3.9163 \times 10^{-13}\,\text{s}$ | $c^2 t^2$ |
| $t_{\varphi_0} = 3.3917 \times 10^{-13}\,\text{s}$ | $c^2 t^2$ |
| $t_{\varsigma P_{Qqd}} = 2.7692 \times 10^{-13}\,\text{s}$ | $c^2 t^2$ |
| $t_{\varsigma 1kg} = 7.2456 \times 10^{-14}\,\text{s}$ | $c^4 t^4$ |
| $t_{\varphi_{E_{Qe}}} = 5.7968 \times 10^{-14}\,\text{s}$ | $c^3 t^2$ |
| $t_{\varphi_{B_{Qe}}} = 5.7968 \times 10^{-14}\,\text{s}$ | $c^2 t^2$ |
| $t_R = 7.0702 \times 10^{-15}\,\text{s}$ | $c^3 t^3 / mol$ |
| $t_{1J} = 4.1847 \times 10^{-18}\,\text{s}$ | $c^6 t^4$ |
| $t_{1C} = 1.7665 \times 10^{-20}\,\text{s}$ | $c^4 t^3$ |
| $t_{\varsigma p} = 1.4658 \times 10^{-20}\,\text{s}$ | $c^4 t^4$ |
| $t_{1N} = 4.5132 \times 10^{-21}\,\text{s}$ | $c^5 t^3$ |
| $t_h = 2.8945 \times 10^{-21}\,\text{s}$ | $c^6 t^5$ |
| $t_{\varsigma e} = 2.2384 \times 10^{-21}\,\text{s}$ | $c^4 t^4$ |
| $t_{h_b} = 2.0048 \times 10^{-21}\,\text{s}$ | $c^6 t^5$ |
| $t_{E_h} = 1.9122 \times 10^{-22}\,\text{s}$ | $c^6 t^4$ |
| $t_{V_P} = 2.2756 \times 10^{-23}\,\text{s}$ | $c^3 t^3$ |
| $t_{G_F} = 3.9986 \times 10^{-23}\,\text{s}$ | $c^9 t^7$ |
| $t_{\varsigma k_B} = 8.3722 \times 10^{-23}\,\text{s}$ | $c^3 t^3 / atom$ |
| $t_{1eV} = 8.3722 \times 10^{-23}\,\text{s}$ | $c^6 t^4$ |
| $t_{1W} = 6.7434 \times 10^{-24}\,\text{s}$ | $c^6 t^3$ |
| $t_{1cd} = 7.6573 \times 10^{-25}\,\text{s}$ | $c^6 t^3 (sr)^{-1}$ |

$$t_{1F} = 3.1477 \times 10^{-25}\text{s} \qquad c^2 t^2$$
$$t_{\mu_B} = 1.5443 \times 10^{-25}\text{s} \qquad c^6 t^4$$
$$t_{\varsigma_{Q_e}} = 9.5943 \times 10^{-27}\text{s} \qquad c^4 t^4$$
$$t_{e'} = 9.5943 \times 10^{-27}\text{s} \qquad c^4 t^3$$
$$t_{1K} = 7.2142 \times 10^{-27}\text{s} \qquad c^3 t$$
$$t_{1A} = 2.3478 \times 10^{-30}\text{s} \qquad c^4 t^2$$
$$t_{V_{k_B}} = 2.2198 \times 10^{-36}\text{s} \qquad c^3 t^3$$
$$t_{V_{Q_e}} = 1.2356 \times 10^{-41}\text{s} \qquad c^3 t^3$$
$$t_P = 5.3912 \times 10^{-44}\text{s} \qquad t$$
$$t_{1Pa} = 8.2624 \times 10^{-45}\text{s} \qquad c^3 t$$
$$t_{1\Omega} = 9.9081 \times 10^{-50}\text{s} \qquad c^{-2} t^{-1}$$
$$t_{k_e''} = 3.3050 \times 10^{-51}\text{s} \qquad c^{-1} t^{-1}$$
$$t_f = 2.6301 \times 10^{-52}\text{s} \qquad c^{-3} t^{-1}$$
$$t_{\varepsilon_0} = 2.6301 \times 10^{-52}\text{s} \qquad ct$$
$$t_{\mu_0} = 2.6301 \times 10^{-52}\text{s} \qquad c^{-3} t^{-1}$$
$$t_{k_e''/4\pi} = 2.6301 \times 10^{-52}\text{s} \qquad c^{-1} t^{-1}$$
$$t_{Z_0} = 2.6301 \times 10^{-52}\text{s} \qquad c^{-2} t^{-1}$$
$$t_{G_0} = 7.6773 \times 10^{-54}\text{s} \qquad c^2 t$$
$$t_{R_K} = 3.8387 \times 10^{-54}\text{s} \qquad c^{-2} t^{-1}$$
$$t_a = 7.0199 \times 10^{-76}\text{s} \qquad ct$$

# Appendix B
## SI Unit to $ct$ Conversions

| | SI derived units | Alternative units * | |
|---|---|---|---|
| $l = ct$ | m | (ct) | (t= 3.3356 x $10^{-9}$ s) |
| $A = (ct)^2$ | m² | (ct)² | (t= 3.3356 x $10^{-9}$ s) |
| $V = (ct)^3$ | m³ | (ct)³ | (t= 3.3356 x $10^{-9}$ s) |
| $m_{m4} = (ct)^4$ | m⁴ | (ct)⁴ | (t= 3.3356 x $10^{-9}$ s) |
| Kilogram: 1 kg = 2.2262 x $10^{-19}$ | m⁴ | (c⁴t⁴) | (t= 7.2456 x $10^{-14}$ s) |
| Newton: 1 N = 2.2262 x $10^{-19}$ | m⁵s⁻² | (c⁵t³) | (t= 4.5132 x $10^{-21}$ s) |
| Joule: 1 J = 1 Nm= 2.2262 x $10^{-19}$ | m⁶s⁻² | (c⁶t⁴) | (t= 4.1847 x $10^{-18}$ s) |
| Ampere: 1 A = 2 x $10^{-7}$ N/m x 2.2262 x $10^{-19}$ = 4.4525 x $10^{-26}$ | m⁴s⁻² | (c⁴t²) | (t= 2.3478 x $10^{-30}$ s) |
| Coulomb: 1 C = 1A x 1 s = 4.4525 x $10^{-26}$ | m⁴s⁻¹ | (c⁴t³) | (t= 1.7665 x $10^{-20}$ s) |
| Volt: 1 V = 1 J/1C = 5.0000 x $10^6$ | m²s⁻¹ | (c²t) | (t= 5.5632 x $10^{-11}$ s) |
| Watt: 1 W = 1 J/s = 2.2262 x $10^{-19}$ | m⁶s⁻³ | (c⁶t³) | (t= 6.7434 x $10^{-24}$ s) |
| Farad: 1 F = 1 C/1V = 8.9050 x $10^{-33}$ | m²s⁻² | (c²t²) | (t= 3.1477 x $10^{-25}$ s) |
| Candela: 1 cd = 3.2595 x $10^{-22}$ | m⁶s⁻³sr⁻¹ | (c⁶t³sr⁻¹) | (t= 7.6573 x $10^{-25}$ s) |
| Kelvin: 1 K = 0.19438 | m³s⁻² | (c³t) | (t= 7.2142 x $10^{-27}$ s) |
| Pascal: 1 Pa = 2.2262 x $10^{-19}$ | m³s⁻² | (c³t) | (t= 8.2624 x $10^{-45}$ s) |
| Electron volt: 1 eV = 1.6022 x $10^{-19}$ J= 3.5669 x $10^{-38}$ | m⁶s⁻² | (c⁶t⁴) | (t= 8.3722 x $10^{-23}$ s) |
| Tesla: 1T = 5 x $10^6$ | (unit-less) | (c⁰t⁰) | |

* The SI unit of mass, the kilogram, as a component of three base units and seventeen derived units, is critical to practically all fields of physics. It was the only SI unit, until recently, not connected to a physical experiment. This theory provides a physical connection relating it to the speed of light. All units can be defined in terms of $c$ and $t$. For example, a joule could be defined in units of $c^6t^4$, a newton is $c^5t^3$, etc. The alternative units can be standardized to current units by specifying $t$. As an example, $1N = 2.2262 \times 10^{-19}$ (c⁵t³). Solving for t, one newton equals $c^5t^3$ when $t$ equals $4.5132 \times 10^{-21}$s.

# Appendix C
## Data: Charge, Electron, Proton, Kilogram, Meter[4]

Physical Elements by $c$ and $t$ Analysis
(Values standardized by volume-time or scat-time for comparison)

$c^7 t^5$

Acsess

Accelerated sess

$|ACS_\varsigma| = |ACS_E| = c^7 t_\varsigma^5$

$c^7 t_{\varsigma Qe}^5 = 1.7693 \times 10^{-71}$ m⁷/s²

$c^7 t_{\varsigma e}^5 = 1.2231 \times 10^{-44}$ m⁷/s²

$c^7 t_{\varsigma p}^5 = 1.4726 \times 10^{-40}$ m⁷/s²

$c^7 t_{\varsigma 1kg}^5 = 4.3462 \times 10^{-7}$ m⁷/s²

$c^7 t_{\varsigma 1m4}^5 = 8.9876 \times 10^{16}$ m⁷/s²

$c^6 t^6$

Sess

Sustained energy space squared

$S^2 = c^6 t_\varsigma^6$:

$S_{Qe}^2 = 5.6625 \times 10^{-106}$ m⁶

$S_e^2 = 9.1324 \times 10^{-74}$ m⁶

$S_p^2 = 7.2003 \times 10^{-69}$ m⁶

$S_{1kg}^2 = 1.0594 \times 10^{-28}$ m⁶

$c^6 t^5$

Moscop

Moving Scop

Planck's constant

$h' = 1.4751 \times 10^{-52}$ m⁶/sψ (not continuous)

$c^6 t_{\varsigma Qe}^5 = 5.9019 \times 10^{-80}$ m⁶/s

$c^6 t_{\varsigma e}^5 = 4.0799 \times 10^{-53}$ m⁶/s

$c^6 t_{\varsigma p}^5 = 4.9122 \times 10^{-49}$ m⁶/s

$c^6 t_{\varsigma 1kg}^5 = 1.4497 \times 10^{-15}$ m⁶/s

$c^6 t_{\varsigma 1m4}^5 = 2.9979 \times 10^{8}$ m⁶/s

$c^6 t^4$

Acscop

Accelerated scop

Accelerating space squared

Energy

$E = V_c c^3 t_{g'} = c^6 t_\varsigma^4$

$E_0 = H_0^2 t_f^2 = 3.4739 \times 10^{-156}$ m⁶/s²

$E_{Qe} = 6.1514 \times 10^{-54}$ m⁶/s²

$E_e = 1.8226 \times 10^{-32}$ m⁶/s²

$E_p = 3.3513 \times 10^{-29}$ m⁶/s²

$E_{1kg} = 2.0008 \times 10^{-2}$ m⁶/s²

$c^5 t^5$

Scop

Sustained consumptive pattern

$c^5 t_{\varsigma Qe}^5 = 1.9687 \times 10^{-88}$ m⁵

$c^5 t_{\varsigma e}^5 = 1.3609 \times 10^{-61}$ m⁵

$c^5 t_{\varsigma p}^5 = 1.6385 \times 10^{-57}$ m⁵

$c^5 t_{\varsigma 1kg}^5 = 4.8357 \times 10^{-24}$ m⁵

$c^5 t_{\varsigma 1m4}^5 = 1$ m⁵



Eric Mitchell Horn

$$E_P = 38.426 \, \text{m}^6/\text{s}^2$$
$$E_{1m4} = 8.9876 \times 10^{16} \, \text{m}^6/\text{s}^2$$

$$c^5 t^4$$

Moscat

Moving scat

Momentum

Maximum momentum, $p_{max}$

$$p_{max} = V_c c^2 t_{g'} = c^5 t_\varsigma^4:$$
$$p_{max_{Qe}} = 2.0519 \times 10^{-62} \, \text{m}^5/\text{s}$$
$$p_{max_e} = 6.0797 \times 10^{-41} \, \text{m}^5/\text{s}$$
$$p_{max_p} = 1.1179 \times 10^{-37} \, \text{m}^5/\text{s}$$
$$p_{max_{1kg}} = 6.6741 \times 10^{-11} \, \text{m}^5/\text{s}$$
$$p_{max_{1m4}} = 2.9979 \times 10^{8} \, \text{m}^5/\text{s}$$

$$c^5 t^3$$

Acscat

Accelerated scat

Accelerating scop

Gravity

$$g' = c^2 V_c = c^5 t_V^3$$
$$g_0 = 2.5796 \times 10^{-183} \, \text{m}^5/\text{s}^2$$
$$g_{Qe} = 4.5681 \times 10^{-81} \, \text{m}^5/\text{s}^2$$
$$g_e = 1.3535 \times 10^{-59} \, \text{m}^5/\text{s}^2$$
$$g_p = 2.4886 \times 10^{-56} \, \text{m}^5/\text{s}^2$$
$$g_{1kg} = 1.4858 \times 10^{-29} \, \text{m}^5/\text{s}^2$$
$$g_{1m4} = 6.6741 \times 10^{-11} \, \text{m}^5/\text{s}^2$$

Force

$$F_{max} = c^5 t_\varsigma^3:$$
$$F_{max_{Qe}} = 2.1387 \times 10^{-36} \, \text{m}^5/\text{s}^2$$
$$F_{max_e} = 2.7160 \times 10^{-20} \, \text{m}^5/\text{s}^2$$
$$F_{max_p} = 7.6263 \times 10^{-18} \, \text{m}^5/\text{s}^2$$
$$F_{max_{1kg}} = 9.2113 \times 10^{2} \, \text{m}^5/\text{s}^2$$
$$F_{max_{1m4}} = 8.9876 \times 10^{16} \, \text{m}^5/\text{s}^2$$

$$c^4 t^4$$

Scat/Mass

$$. \; m_{m4} = \varsigma = V_c c t_g = c^4 t_\varsigma^4$$
$$\varsigma_0 = 3.8652 \times 10^{-173} \, \text{m}^4$$
$$\varsigma_{Qe} = 6.8444 \times 10^{-71} \, \text{m}^4$$
$$\varsigma_e = 2.0280 \times 10^{-49} \, \text{m}^4$$
$$\varsigma_p = 3.7288 \times 10^{-46} \, \text{m}^4$$
$$\varsigma_{1kg} = 2.2262 \times 10^{-19} \, \text{m}^4$$
$$m_{1m4} = c^4 t_{\varsigma_{1m4}}^4 = 1 \, \text{m}^4$$

$$c^4 t^3$$

Moving space

Moving volume

$$mov = \frac{\varsigma}{t_{g'}} = c V_c = c^4 t_V^3$$

$$mo\,v_0 = 8.6046 \times 10^{-192}\,\text{m}^4\text{/s}$$

$$cV_{c_{Qe}} = 1.5238 \times 10^{-89}\ \text{m}^4\text{/s}$$
$$cV_{c_e} = 4.5147 \times 10^{-68}\ \text{m}^4\text{/s}$$
$$cV_{c_p} = 8.3011 \times 10^{-65}\ \text{m}^4\text{/s}$$
$$cV_{c_{1kg}} = 4.9561 \times 10^{-38}\ \text{m}^4\text{/s}$$
$$cV_{c_{1m4}} = 2.2262 \times 10^{-19}\,\text{m}^4\text{/s}$$

Most
Moving space with time
Charge

$$most = cV_\varsigma = c^4 t_\varsigma^3:$$
$$Q_e = c^4 t_{\varsigma_{Qe}}^3 = 7.1338 \times 10^{-45}\ \text{m}^4\text{/s}$$
$$c^4 t_{\varsigma_e}^3 = 9.0597 \times 10^{-29}\ \text{m}^4\text{/s}$$
$$c^4 t_{\varsigma_p}^3 = 2.5439 \times 10^{-26}\ \text{m}^4\text{/s}$$
$$c^4 t_{\varsigma_{1kg}}^3 = 3.0725 \times 10^{-6}\ \text{m}^4\text{/s}$$
$$c^4 t_{\varsigma_{1m4}}^3 = 2.9979 \times 10^8\ \text{m}^4\text{/s}$$

$$c^4 t^2$$

Acsp
Accelerated space
Accelerating scat/mass
$$\varsigma_{a'} = c^4 t_\varsigma^2$$
$$\varsigma_{a'_{Qe}} = 7.4355 \times 10^{-19}\ \text{m}^4\text{/s}^2$$
$$\varsigma_{a'_e} = 4.0474 \times 10^{-8}\ \text{m}^4\text{/s}^2$$
$$\varsigma_{a'_p} = 1.7355 \times 10^{-6}\ \text{m}^4\text{/s}^2$$
$$\varsigma_{a'_{1kg}} = 4.2406 \times 10^7\ \text{m}^4\text{/s}^2$$
$$\varsigma_{a'_{1m4}} = 8.9876 \times 10^{16}\ \text{m}^4\text{/s}^2$$

$$c^3 t^3$$

Space
$$S = V_S = c^3 t^3$$
Consumptive volume
$$V_c = (c\,t_V)^3:$$
$$S_0 = 2.8702 \times 10^{-200}\,\text{m}^3$$
$$V_{c_{Qe}} = 5.0827 \times 10^{-98}\,\text{m}^3$$
$$V_{c_e} = 1.5059 \times 10^{-76}\,\text{m}^3$$
$$V_{c_p} = 2.7690 \times 10^{-73}\,\text{m}^3$$
$$V_{c_{1kg}} = 1.6532 \times 10^{-46}\,\text{m}^3$$
$$V_{c_{1m4}} = 7.4259 \times 10^{-28}\,\text{m}^3$$

Scat-space

$$S = (c\,t_\varsigma)^3$$
$$S_{Qe} = 2.3796 \times 10^{-53}\,\text{m}^3$$
$$S_{\varsigma_e} = 3.0220 \times 10^{-37}\,\text{m}^3$$
$$S_{\varsigma_p} = 8.4854 \times 10^{-35}\,\text{m}^3$$
$$S_{\varsigma_{1kg}} = 1.0249 \times 10^{-14}\,\text{m}^3$$
$$S_{\varsigma_{1m4}} = 1\ \text{m}^3$$



*Eric Mitchell Horn*

$$c^3 t^2$$

Moar

Moving Area

moving voar (mova)*

$$mova = c^3 t_V^2$$

$$mova_{Q_e} = 4.1135 \times 10^{-57} \text{m}^3\text{/s}$$

$$mova_e = 8.4859 \times 10^{-43} \text{m}^3\text{/s}$$

$$mova_p = 1.2736 \times 10^{-40} \text{m}^3\text{/s}$$

$$mova_{1kg} = 9.0303 \times 10^{-23} \text{m}^3\text{/s}$$

$$mova_{1m4} = 2.4584 \times 10^{-10} \text{m}^3\text{/s}$$

moving scat-area (mosa)

$$mosa = c^3 t_\varsigma^2$$

$$\sqrt{E}, c\sqrt{\varsigma}, cA_\varsigma, \frac{V_\varsigma}{t_\varsigma}, c^3 t_\varsigma^2:$$

$$\sqrt{E_{Qe}} = 2.4802 \times 10^{-27} \text{m}^3\text{/s}$$

$$\sqrt{E_e} = 1.3501 \times 10^{-16} \text{m}^3\text{/s}$$

$$\sqrt{E_p} = 5.7890 \times 10^{-15} \text{m}^3\text{/s}$$

$$\sqrt{E_{1kg}} = 0.14145 \text{ m}^3\text{/s}$$

$$\sqrt{E_{1m4}} = 2.9979 \times 10^8 \text{m}^3\text{/s}$$

$$c^3 t$$

Acar

Accelerated area

$$Acar = A\frac{c}{t}$$

Accelerating space due to scat/mass

$$S_{a_\varsigma'} = c^3 t_\varsigma$$

$$H_0 = c^3 t_f = 7.0865 \times 10^{-27} \text{m}^3\text{/s}^2$$

$$S_{a'_{\varsigma Qe}} = 0.2585 \text{ m}^3\text{/s}^2$$

$$S_{a'_{\varsigma e}} = 6.0312 \times 10^4 \text{ m}^3\text{/s}^2$$

$$S_{a'_{\varsigma p}} = 3.9494 \times 10^5 \text{ m}^3\text{/s}^2$$

$$S_{a'_{\varsigma 1kg}} = 1.9522 \times 10^{12} \text{ m}^3\text{/s}^2$$

$$S_{a'_{\varsigma P}} = 1.2924 \times 10^{13} \text{ m}^3\text{/s}^2$$

$$S_{a'_{\varsigma 1m4}} = 8.9876 \times 10^{16} \text{ m}^3\text{/s}^2$$

* May not be theoretically relevant

$$c^2 t^2$$

Area

$$A = c^2 t^2$$

Voar (volume-area)

$$A_V = c^2 t_V^2$$

Scar (scat-area)

$$A_\varsigma = (ct_\varsigma)^2$$

$$A_{\varsigma Qe} - 8.2731 \times 10^{-36} \text{m}^2$$

$$A_{\varsigma e} = 4.5033 \times 10^{-25} \text{m}^2$$

$$A_{\varsigma p} = 1.9310 \times 10^{-23} \text{m}^2$$

$$A_{\varsigma 1kg} = 4.7183 \times 10^{-10} \text{m}^2$$

$$A_{\varsigma P} = 2.0677 \times 10^{-8} \text{m}^2$$

$$A_{\varsigma 1m4} = 1 \text{m}^2$$

$$c^2 t$$

Moving distance

$$modi = c^2 t$$

$$G_0 = 4\alpha c^2 t_f = 6.9000 \times 10^{-37}\,\text{m}^2/\text{s}$$

$$R_K'^{-1} = 2\alpha c^2 t_f = 3.4500 \times 10^{-37}\,\text{m}^2/\text{s}$$

$$Z_0^{-1} = c^2 t_f = 2.3638 \times 10^{-35}\,\text{m}^2/\text{s}$$

$$c^2 t_{\varsigma Qe} = 8.6229 \times 10^{-10}\,\text{m}^2/\text{s}$$

$$c^2 t_{\varsigma e} = 2.0118 \times 10^{-4}\,\text{m}^2/\text{s}$$

$$c^2 t_{\varsigma p} = 1.3174 \times 10^{-3}\,\text{m}^2/\text{s}$$

$$c^2 t_{\varsigma 1kg} = 6.5120 \times 10^{3}\,\text{m}^2/\text{s}$$

$$c^2 t_{\varsigma P} = 4.3109 \times 10^{4}\,\text{m}^2/\text{s}$$

$$c^2 t_{\varsigma 1m4} = 2.9979 \times 10^{8}\,\text{m}^2/\text{s}$$

$$c^2$$

Acdi

Accelerated distance

$$c^2 = \frac{c}{t} * ct$$

Accelerating area

$$A_{a'} = nc^2$$

$$ct$$

Distance

$$d = ct$$

Volume length

$$l_V = ct_V:$$

$$a_0 = 2.1045 \times 10^{-67}\,\text{m}$$

$$ct_{V Qe} = 3.7042 \times 10^{-33}\,\text{m}$$

$$ct_{V e} = 5.3203 \times 10^{-26}\,\text{m}$$

$$ct_{V p} = 6.5179 \times 10^{-25}\,\text{m}$$

$$ct_{V 1kg} = 5.4883 \times 10^{-16}\,\text{m}$$

$$ct_{V P} = 6.8221 \times 10^{-15}\,\text{m}$$

$$ct_{V 1m4} = 9.0556 \times 10^{-10}\,\text{m}$$

Scat-length

$$l_\varsigma = ct_\varsigma$$

$$\varepsilon_0' = ct_f = 7.8851 \times 10^{-44}\,\text{m}$$

$$ct_{\varsigma Qe} = 2.8763 \times 10^{-18}\,\text{m}$$

$$ct_{\varsigma e} = 6.7107 \times 10^{-13}\,\text{m}$$

$$ct_{\varsigma p} = 4.3943 \times 10^{-12}\,\text{m}$$

$$ct_{\varsigma 1kg} = 2.1722 \times 10^{-5}\,\text{m}$$

$$ct_{\varsigma P} = 1.4380 \times 10^{4}\,\text{m}$$

$$ct_{\varsigma 1m4} = 1\,\text{m}$$



*Eric Mitchell Horn*

Time

$$t_{R_K} = 3.8387 \times 10^{-54}\text{s}$$

$$t_0 = \frac{a_0}{c} = 7.0199 \times 10^{-76}\text{s}$$

Volume-time

$$t_V = (\frac{V_c}{c^3})^{\frac{1}{3}}$$

$$t_{V_{Qe}} = 1.2356 \times 10^{-41}\text{s}$$
$$t_{V_e} = 1.7747 \times 10^{-34}\text{s}$$
$$t_{V_p} = 2.1741 \times 10^{-33}\text{s}$$
$$t_{V_{1kg}} = 1.8307 \times 10^{-24}\text{s}$$
$$t_{V_P} = 2.2756 \times 10^{-23}\text{s}$$
$$t_{V_{1m4}} = 3.0206 \times 10^{-18}\text{s}$$

Scat-time

$$t_\varsigma = (\frac{m_{m4}}{c^4})^{\frac{1}{4}}$$
$$t_f = 2.6301 \times 10^{-52}\text{s}$$
$$t_{\varsigma_{Qe}} = 9.5943 \times 10^{-27}\text{s}$$
$$t_{\varsigma_e} = 2.2384 \times 10^{-21}\text{s}$$
$$t_{\varsigma_p} = 1.4658 \times 10^{-20}\text{s}$$
$$t_{\varsigma_{1kg}} = 7.2456 \times 10^{-14}\text{s}$$
$$t_{\varsigma_P} = 4.7965 \times 10^{-13}\text{s}$$
$$t_{\varsigma_{1m4}} = 3.3356 \times 10^{-9}\text{s}$$

# Kilogram Statistics and Conversions

*Conversion between a mass in kilograms and a scat/mass in m⁴:*

$$\varsigma = m_{m^4} = m_{kg} * 2.2262 \times 10^{-19} \; (m^4)$$

*The volume-time of a kilogram and the volume-time of a mass expressed in kilograms, $t_V$:*

$$m_{m^4} = \varsigma = V_c c t_g = (c t_V)^3 c t_g$$

$$t_{V_{1kg}} = \left(\frac{\varsigma}{c^4 t_g}\right)^{\frac{1}{3}} = 1.8307 \times 10^{-24} \text{s}$$

$$t_V = (m_{kg})^{\frac{1}{3}} * 1.8307 \times 10^{-24} \; (s)$$

*The scat-time of a kilogram, $t_{\varsigma 1kg}$, and the scat-time of a mass expressed in kg, $t_\varsigma$:*

$$m_{m^4} = c^4 t_\varsigma^4$$

$$t_{\varsigma 1kg} = \left(\frac{\varsigma_{1kg}}{c^4}\right)^{\frac{1}{4}} = \left(\frac{m_{1kg} * 2.2262 \times 10^{-19}}{c^4}\right)^{\frac{1}{4}} = 7.2456 \times 10^{-14} \text{s}$$

$$t_{\varsigma kg} = m_{kg}^{\frac{1}{4}} * 7.2456 \times 10^{-14} \; (s)$$

*The voar (volume-area) of a kilogram, $A_{V_{1kg}}$, and the voar of a mass expressed in kg, $A_V$ $(c t_V)^2$:*

$$A_{V1kg} = \frac{A_{\varsigma 1kg}^2}{c^2 t_{V1kg} t_g} = \frac{V_{c1kg}}{c t_{V1kg}} = \frac{1.6532 \times 10^{-46}}{c * 1.8307 \times 10^{-24}} = 3.0122 \times 10^{-31} \; m^2$$

$$A_V = m_{kg}^{\frac{2}{3}} * 3.0122 \times 10^{-31} \; (m^2)$$



*Eric Mitchell Horn*

The scat-area of a kilogram, $A_{\varsigma 1kg}$ and the scat-area of a mass expressed in kg , $A_\varsigma$, $(ct_\varsigma)^2$:

$$A_{\varsigma 1kg} = \sqrt{\varsigma_{1kg}} = \sqrt{2.2262 \times 10^{-19}} = 4.7183 \times 10^{-10} \text{ m}^2$$

$$A_\varsigma = \sqrt{m_{kg}} * 4.7183 \times 10^{-10} \text{ (m}^2\text{)}$$

The consumptive volume of a kilogram, $V_{c1kg}$, and the consumptive volume of a mass expressed in kg, $V_c$, $(ct_V)^3$ :

$$V_c = \frac{\varsigma}{ct_g}$$

$$V_{c1kg} = \frac{m_{1kg} * 2.2262 \times 10^{-19}}{ct_{g'}} = 1.6532 \times 10^{-46} \text{ m}^3$$

$$V_c = m_{kg} * 1.6532 \times 10^{-46} \text{ (m}^3\text{)}$$

The scat-volume of a kilogram, $V_{\varsigma 1kg}$, and the scat-volume of a mass expressed in kg, $(ct_\varsigma)^3$:

$$V_{\varsigma 1kg} = (c * 7.2456 \times 10^{-14})^3 = 1.0249 \times 10^{-14} \text{ m}^3$$

$$V_\varsigma = m_{kg}^{\frac{3}{4}} * 1.0249 \times 10^{-14} \text{ (m}^3\text{)}$$

Mova, the moving voar of a kilogram, and the moving voar of a mass expressed in kg, $c^3 t_V^2$ :

$$mova = cA_V$$

$$mova_{1kg} = cA_{c1kg} = c * 3.0121 \times 10^{-31} = 9.0300 \times 10^{-23} \text{m}^3\text{/s}$$

$$movava_{kg} = cA_{c_{kg}} = m_{kg}^{\frac{2}{3}} * 9.0300 \times 10^{-23} \ \text{(m³/s)}$$

*Mosa, the moving scat-area of a kilogram and the moving scat-area of a mass expressed in kg, $\sqrt{E}$, $(c^3 t_c^2)$:*

$$\frac{V_{c1kg}}{t_c} = \sqrt{E_{1kg}} = c\sqrt{c_{1kg}} = cA_{c1kg} = c\sqrt{2.2262 \times 10^{-19}} = c * 4.7183 \times 10^{-10}$$

$$mosa_{1kg} = 0.14145 \ \text{m³s}^{-1}$$

$$mosa_{kg} = \frac{V_c}{t_c} = \sqrt{m_{kg}} * 0.14145 \ \text{(m³/s)}$$

*The acceleration of space caused by a scat/mass of 1 kilogram, $S_{a_{c1kg}}$, and a scat/mass expressed in kg, $S_{a_c}$, $(c^3 t_c)$:*

$$S_{a_c} = \frac{\sqrt{E}}{t_c} = \frac{cA_c}{t_c} = c^3 t_c$$

$$S_{a_{c1kg}} = \frac{cA_{c1kg}}{t_{c1kg}} = \frac{\sqrt{E_{1kg}}}{t_{c1kg}} = \frac{0.14145}{7.2456 \times 10^{-14}} = 1.9522 \times 10^{12} \ \text{m³/s}^2$$

$$S_{a_c} = (m_{kg})^{\frac{1}{4}} * 1.9522 \times 10^{12} \ \text{(m³/s}^2)$$

*The temperature loss due to a 1 kilogram scat/mass, $T_{loss_{g1kg}}$:*

$$T_{loss_{g1kg}} = c^3 t_{c1kg} = 1.9522 \times 10^{12} \ \text{m³/s}^2 = 3.7948 \times 10^{11} \text{K}$$

*The gravitational force of 1 kilogram and the gravitational force of scat/mass expressed in kg, $c^2 V_c$, $c^5 t_V^3$:*

$$g_{1kg} = c^2 V_{c1kg} = c^2 * 1.6532 \times 10^{-46} = 1.4858 \times 10^{-29} \ \text{m⁵/s}^2$$



*Eric Mitchell Horn*

$$g' = m_{kg} * 1.4858 \times 10^{-29} \ (\text{m}^5/\text{s}^2)$$

*The energy of 1 kilogram, $E_{1kg}$, and the energy of a scat/mass expressed in kg, $c^6 t_{\varsigma}^4$:*

$$E_{1kg} = \frac{V_S^2}{t^2} = (cA_{\varsigma 1kg})^2 = 2.0008 \times 10^{-2} \ \text{m}^6\text{s}^{-2} = 8.9876 \times 10^{16} \text{J}$$

$$E_{\varsigma kg} = (m_{kg}) * 0.020008 \ (\text{m}^6/\text{s}^2)$$

# Appendix E
## Selected Other Quantum Constant Conversions

The following work needs review by professionals and this work remains unfinished.

*The electromagnetic force quantum of electron charge, (hypothesized):*

$$F_{max_{Q_e}} = Q_e * c = c^5 t_{\varsigma_{Q_e}}^3 = c * c^2 t_f * \varphi_{B_{Q_e}} = c^3 t_f \varphi_{B_{Q_e}} = H_0 \varphi_{B_{Q_e}}$$

$$k_{Q_e F} = c$$

$$k_{Q_e F} = k_{e''} * c^2 t_{k_e}$$

*The strong force quantum of electromagnetic charge, (hypothesized):*

The strong force is approximately 137 X the electromagnetic force. Using this:

$$\frac{1}{\alpha} \approx 137$$

$$F_{strong_{Q_e}} = F_{Q_e} * \frac{2t_f}{t_{R_K}} = \frac{2c^3 t_f^2 * \varphi_{B_{Q_e}}}{t_{R_K}}$$

*The nuclear magneton constant, $\mu_N$:*

$$\mu_N = \frac{Q_e h_b}{2m_P} = 5.0508 \times 10^{-27} \text{J/T}$$

$$= \frac{(5.0508 \times 10^{-27})(2.2262 \times 10^{-19})}{(5 \times 10^6)} = 2.2488 \times 10^{-52} \text{m}^6/\text{s}^2$$

$$= c^6 t^4 \text{ when } t = 2.3592 \times 10^{-26} \text{s}$$

i

$$t_{\mu_N} = 2.3592 \times 10^{-26} \text{s}$$

*The Bohr magneton constant, $\mu_B$:*



Eric Mitchell Horn

$$\mu'_B = \frac{9.2740 \times 10^{-24}}{5 \times 10^6} \text{ J/T}$$

$$= \frac{(9.274 \times 10^{-24})c^6 t^4_{1J}}{5 \times 10^6} = \frac{c^6(9.274 \times 10^{-24})(4.1847 \times 10^{-18})^4}{5 \times 10^6} = 4.1292 \times 10^{-49}$$

m⁶/s²

$\mu'_B = 4.1292 \times 10^{-49}$ m⁶/s² $= c^6 t^4$ when $t^4 = 5.6878 \times 10^{-100}$ or
$t = 1.5443 \times 10^{-25}$s

$$t_{\mu_B} = 1.5443 \times 10^{-25}\text{s}$$

Checking:

$$\mu'_B = \frac{e'h'_b}{2\varsigma_e} = \frac{(7.1338 \times 10^{-45})(2.3477 \times 10^{-53})}{2(2.0280 \times 10^{-49})} = 4.1292 \times 10^{-49}\text{m⁶/s²}$$

*Classical electron radius, $r_e$:*

$$r_e = \frac{e^2}{4\pi\varepsilon_0 m_e c^2} = 2.8179 \times 10^{-15}\text{m} = ct \text{ when } t = 9.3995 \times 10^{-24}\text{s}$$

$$t_{r_e} = 9.3995 \times 10^{-24}\text{s}$$

*Thomson cross section, $(\frac{8\pi}{3})r_e^2$:*

$$(\frac{8\pi}{3})r_e^2 = 6.6525 \times 10^{-29}\text{m²} = c^2 t^2 \text{ when } t = 2.7206 \times 10^{-23}\text{s}$$

*Hartree energy, $E_h$:*

$$E_h = 2R_\infty hc = 4.3597 \times 10^{-18}\text{J}$$
$$= (4.3597 \times 10^{-18})(2.2262 \times 10^{-19}) = 9.7057 \times 10^{-37} \text{ m⁶/s²}$$

$$= c^6 t^4 \text{ when } t = 1.9122 \times 10^{-22}\text{s}$$

$$t_{E_h} = 1.9122 \times 10^{-22}\text{s}$$

*Fermi coupling constant, $G_F^0$:*

$$G_F^0 = \frac{G_F}{(h_b c)^3} = 1.1664 \times 10^{-5} \text{GeV}^{-2} = \frac{1.1664 \times 10^{-23}}{c^{12} t_{k_B}^8}$$

Performing $ct$ analysis; for this to be correct:

$$h_b = c^6 t_{h_b}^5$$

$$(h_b c)^3 = c^{21} t_{h_b}^{15} = \left(\frac{cS_{Q_e}^2}{2\pi t_{R_K}}\right)^3 = \frac{c^{21} t_{\varsigma_{Q_e}}^{18}}{8\pi^3 t_{R_K}^3}$$

$$G_F^0 = \frac{8\pi^3 t_{R_K}^3 G_F}{c^{21} t_{\varsigma_{Q_e}}^{18}} = \frac{1.1664 \times 10^{-23}}{c^{12} t_{k_B}^8}$$

$$G_F = \frac{(1.1664 \times 10^{-23}) c^{21} t_{\varsigma_{Q_e}}^{18}}{8\pi^3 c^{12} t_{k_B}^8 t_{R_K}^3} = c^9 t_{G_F}^7$$

$$G_F = 3.1969 \times 10^{-81} \text{m}^9/\text{s}^2 = c^9 t_{G_F}^7$$

$$t_{G_F} = \left(\frac{G_F}{c^9}\right)^{\frac{1}{7}}$$

$$t_{G_F} = 3.9986 \times 10^{-23} \text{s}$$

For the final units to have a denominator equal to $GeV^2 = (10^9 c^6 t_{k_B}^4)^2 = 10^{18} c^{12} t_{k_B}^8$, the numerator, $G_F$, must have $ct$ units equal to $c^9 t^7$; i.e. units of accelerated (scat/mass squared); $\frac{c}{t} * c^8 t^8 = c^9 t^7$.

This implies that there are two factors of the gravitational constant of one mass, $k_{g'}$, hidden within the Fermi coupling constant. Let $E_P$ symbolize the maximum energy associated with Planck's constant. This maximum energy has a scat-time, $t_{\varsigma_P}$ and a volume-time, $t_{V_P}$, associated with it.



*Eric Mitchell Horn*

$$E_P = \frac{S_{Q_e}^2}{t_{R_K}^2 \psi}$$

$$E_P = c^6 t_{\varsigma P}^4 = V_{c_P} c^3 t_{g'}$$

$$V_{c_P} = \frac{c^6 t_{\varsigma P}^4}{c^3 t_{g'}} = \frac{c^3 t_{\varsigma P}^4}{t_{g'}}$$

Since Boltzmann's constant is a volume, it will also have an equivalent scat-time, $t_{\varsigma k_B}$ associated with it. Assuming $t_{R_K}$ represents the time associated with a minimum distance, $c t_{R_K}$ and a minimum volume, $c^3 t_{V_{R_K}}^3$ :

$$t_{g'} = \frac{c^3 t_{\varsigma P}^4}{V_{c_P}} = \frac{c^3 t_{\varsigma k_B}^4}{V_{c_{k_B}}} = \frac{c^3 t_{\varsigma Q_e}^4}{V_{c_{Q_e}}} = \frac{c^3 t_{\varsigma R_K}^4}{V_{c_{R_K}}}$$

$$t_{g'} = \frac{t_{\varsigma P}^4}{t_{V_P}^3} = \frac{t_{\varsigma k_B}^4}{t_{V_{k_B}}^3} = \frac{t_{\varsigma Q_e}^4}{t_{V_{Q_e}}^3} = \frac{t_{\varsigma R_K}^4}{t_{V_{R_K}}^3}$$

$$t_{\varsigma k_B}^8 = \frac{t_{V_{k_B}}^6 t_{\varsigma R_K}^8}{t_{V_{R_K}}^6}$$

Showing the two factors of $k_{g'}$:

$$G_F = \frac{1.1664 \times 10^{-23} c^{21} t_{\varsigma Q_e}^{18} t_{V_{R_K}}^3}{(2\pi)^3 c^{12} t_{V_{k_B}}^6 t_{\varsigma R_K}^8} = \frac{1.1664 \times 10^{-23} c^{21} t_{\varsigma Q_e}^{18}}{(2\pi)^3 c^{12} t_{V_{k_B}}^6 t_{V_{R_K}}^3 t_{g'}^2}$$

The above expression possibly hints of another factor of $\dfrac{1}{t_{g'}}$ hidden within the number 1.1664 x 10⁻²³.

$$G_F = \frac{1.1664 \times 10^{-23} c^9 t_{\varsigma P}^{12} t_{V_{R_K}}^3}{8\pi^3 t_{\varsigma k_B}^8} = c^9 t^7$$

Expressing $G_F$ in terms of moving volumes of space:

$$\pi = \frac{t_{k_e}}{4t_f}$$

$$G_F = \frac{(9.3312 \times 10^{-23})c^6 V_{c_P}^3 V_{c_{H_0}} V_{c_{R_K}} t_{g'}^3}{c^3 V_{c_{k_B}}^2 V_{c_{k_e}} t_{g'}^2}$$

This expression has the general form of $\dfrac{c}{t_{g'}}(c^2 V_c^2 t_{g'}^2) = \dfrac{c}{t_{g'}}\varsigma^2$, but there are missing terms hidden within the numeric expression. However, this expression directly connects gravity to $G_F^0$.

Recall:

$$g = c^2 V_c$$

$9.3312 \times 10^{-23}$ is likely a mixed ratio of forces and time elements,

$$\frac{t_1}{t_2}, \frac{t_1^2}{t_2^2}, \frac{t_1^3}{t_2^3}, \frac{t_1 * t_2}{t_3 * t_4}....\text{etc...}$$

The question is one of the relationship between $t_{k_B}$ and $t_{\varsigma P}$, $t_f$, $t_{R_K}$, and $t_{g'}$.

Let $n = 1.1664 \times 10^{-23}$

$$t_{k_B}^4 = t_{V_{k_B}}^3 t_{g'}$$

$$t_{k_B}^8 = t_{V_{k_B}}^6 t_{g'}^2$$

$$t_{\varsigma P}^4 = t_{V_P}^3 t_{g'}$$

$$G_F = \frac{c}{t_{g'}}\varsigma^2$$

$$\varsigma^2 = \frac{n c^8 t_{\varsigma P}^8 t_{V_P}^3 t_{R_K}^3}{8\pi^3 t_{V_{k_B}}^6}$$

$$\varsigma = \frac{c^4 t_{\varsigma P}^4}{t_{V_{k_B}}^3} \sqrt{\frac{n t_{V_P}^3 t_{R_K}^3}{2\pi}}$$

To be continued....

*Quantum of circulation,* $\dfrac{h}{2m_e}$:

$$\frac{h}{2m_e} = 3.6369 \times 10^{-4} \text{m}^2\text{/s} = c^2 t \text{ when } t = 4.0466 \times 10^{-21}\text{s}$$

*The Z boson (average):*

$$E_Z = 9.119 \times 10^{10}\text{eV} \times 3.5669 \times 10^{-38}\text{m}^6\text{/s}^2\text{·eV} = 3.2527 \times 10^{-27}\text{m}^6\text{/s}^2$$

$$\varsigma_Z = 3.6191 \times 10^{-44}\text{m}^4$$

$$mosp_Z = 8.0567 \times 10^{-63}\text{m}^4\text{/s}$$

$$most_Z = 7.8665 \times 10^{-25}\text{m}^4\text{/s}$$

$$V_{c_Z} = 2.6874 \times 10^{-71}\text{m}^3$$

$$V_{\varsigma_Z} = 2.6240 \times 10^{-33}\text{m}^3$$

$$g_Z = 2.4153 \times 10^{-54}\text{m}^5\text{/s}^2$$

*The W boson, (average):*

$$E_W = 80.39 \times 10^9\text{eV} \times 3.5669 \times 10^{-38}\text{m}^6\text{/s}^2\text{·eV} = 2.8674 \times 10^{-27}\text{m}^6\text{/s}^2$$

$$\varsigma_W = 3.1904 \times 10^{-44}\text{m}^4$$

$$mosp_W = 7.1025 \times 10^{-63}\text{m}^4\text{/s}$$

$$most_W = 7.1565 \times 10^{-25}\text{m}^4\text{/s}$$

$$V_{c_W} = 2.3691 \times 10^{-71}\text{m}^3$$

$$V_{\varsigma W} = 2.3872 \times 10^{-33} m^3$$

$$g_W = 2.1293 \times 10^{-54} m^5/s^2$$

$$t_{V_W} = 9.5802 \times 10^{-33} s$$

$$t_{\varsigma W} = 4.4580 \times 10^{-20} s$$

*Wien wavelength displacement law constant, $b$:*

$$b = 2.8978 \times 10^{-3} m \bullet K = (2.8978 \times 10^{-3})(0.19438) = 5.6327 \times 10^{-4} m^4/s^2$$
$$= c^4 t^2 \text{ when } t = 2.6407 \times 10^{-19} s$$

$$t_b = 2.6407 \times 10^{-19} s$$

*Bohr radius, $a_0$:*

$$a_0 = \frac{h_b}{\alpha m_e c} = 5.2919 \times 10^{-11} m = ct \text{ when } t = 1.7652 \times 10^{-19} s$$

$$t_{a_0} = 1.7652 \times 10^{-19} s$$

*Rydberg constant:*

$$R_\infty = 1.0974 \times 10^7 m^{-1} = c^{-1} t^{-1} \text{ when } t = 3.0396 \times 10^{-16} s$$

$$R_\infty = \frac{1}{c t_{R_\infty}}$$

$$t_{R_\infty} = 3.0396 \times 10^{-16} s$$

*Stephan-Boltzmann constant:*

$$\sigma = 5.6704 \times 10^{-8} W/m^2 \bullet K^4 = \frac{(5.6704 \times 10^{-8})(2.2262 \times 10^{-19})}{1 \times (0.19438)^4} s^5/m^8 = c^{-8} t^{-3}$$

when $t = 1.2012 \times 10^{-15} s$

$$\sigma = \frac{1}{c^8 t_\sigma^3} \qquad t_\sigma = 1.2012 \times 10^{-15}$$

# Appendix F
## Summary of Foundational Hypotheses

**Hypothesis:** *Physical phenomena have physical explanations.*

**Hypothesis:** *Space is a quantifiable substance.*

**Hypothesis:** *Space can be divided into parts.*

**Hypothesis:** *Space exists as a particle with a specific size and shape. A space particle, (acronym spart) is the only true particle, the only true point particle, and has no mass.*

**Hypothesis:** *Empty space does not exist. Where there are no space particles, there is no space.*

**Hypothesis:** *The space particle is the smallest thing possible in the universe.*

**Hypothesis:** *The space particle must possess the property of tessellation, the ability to fill space completely without any gaps between the space particles.*

**Hypothesis:** *Individual space particles are undetectable.*

**Hypothesis:** *The universe consists of an observable universe, henceforth referred to as the outer-verse, designated $U_o$, and an at present, unobservable part of the universe, henceforth referred to as the inner-verse, designated $U_i$. In mathematical terms, the total universe, $U_t$ is equal to the sum of the outer-verse plus the inner-verse:*

$$U_t = U_o + U_i$$

**Hypothesis:** *A boundary exists between the outer-verse and the inner-verse.*

**Hypothesis:** *Multiple unobserved Inner-verses may exist. This is a subject for future consideration.*

**Hypothesis:** *The space particle is the currency of exchange in the universe.*

**Hypothesis:** *The outer-verse, $U_o$, and the inner-verse, $U_i$, are connected physically to one another via hole-complexes.*

**Hypothesis:** *Space particles flow through a network of hole-complexes to and from the inner-verse.*

**Hypothesis:** *The subset of gravity results from a non-directionalized consumption of space particles. (Space particles are consumed equally from all directions. The direction of consumption can be controlled by electromagnetic means.)*

**Hypothesis:** *Once consumed, space particles continue to exist in an unobservable part of the universe, $U_i$. Space particles are not crushed, digested to nothingness, nor eroded by time.*

**Hypothesis:** *Space in the outer-verse is continually being replaced.*

**Hypothesis:** *Individual space particles do not travel through space. Only holes or hole-complexes which connect the outer-verse to the inner-verse travel through space.*

**Hypothesis:** *Individual space particles only move with space particle flows.*

**Hypothesis:** *Space particles travel at the speed of light when passing from the outer-verse to the inner-verse and vice versa.*

**Hypothesis:** *Matter, an object with the property of mass, is touchable moving space through time.*

**Hypothesis:** *The difference between the space we can touch; i.e. ordinary matter, and the space previously referred to empty space is the velocity of the space particle flow at the touch point.*

**Hypothesis:** *Space is the equivalent of a flowing solid. The flow of space is caused by the consumption of space by an object with mass with the simultaneous replacement of space by energy.*

**Hypothesis:** *Matter is formed from the flow of space particles which are traveling near the speed of light as the space particles approach the hole complexes, and the interactions of these space particle flows.*

**Hypothesis:** *Matter can be quantified from several equally valid perspectives, the net area of the consumptive hole-complex, $A_\varsigma$; the net volume of space that is continuously traveling at the speed of light through the holes, $V_c$, or the volume of space consumed in a length of time specifically associated with an object with mass, $V_\varsigma$, or scat/rest mass, $\varsigma$.*

**Hypothesis:** *Space multiplied by the speed of light and time (acronym **scat**) is an equation for rest mass.*

**Hypothesis:** *The property of mass results from a net volume of space, $V_c$ ( consumptive volume), traveling at the speed of light through a hole complex as long as the hole-complex exists.*

**Hypothesis:** *The property of mass is a poor quantifier of matter since it changes with velocity, gravitation, and temperature.*

**Hypothesis:** *The property of mass measures the energy required to change a particular consumptive pattern.*

**Hypothesis:** *The property of mass and the quantity of rest mass are two different entities; i.e.* $m_{relativity} \neq m_{restmass}$

**Hypothesis:** *When viewing rest mass as a quantity of matter, an object possessing the property of mass can be represented by the following equation:*

$$matter = restmass = m_{m^4} = scat = V_c * c * t.$$

$m_{m^4}$ = *mass expressed in meters raised to the 4th power. $V_c$ is the consumptive volume; the volume of space continuously traveling at the speed of light that forms matter.*

**Hypothesis:** *Rest mass does not change with velocity or gravitation.*

**Hypothesis:** *The consumptive pattern of the hole-complex changes relative to velocity or space particle flows resulting from a gravitational process.*

**Hypothesis:** *All physical constants can be expressed as ratios between various combinations of powers of c, the speed of light; $t_\varsigma$, the scat-time; $t_V$, the volume-time associated with an object with mass, and other fundamental time constants and time ratios.*

**Hypothesis:** *All objects composed of matter are attracted to all other objects composed of matter via the consumption of the media surrounding matter.*

**Hypothesis:** *The media surrounding matter is space.*

**Hypothesis:** *Directionalized gravity is a net proportioned directionalized consumption of space particles.*

**Hypothesis:** *A sustained directionalized consumptive pattern results in movement of matter.*

**Hypothesis:** *Forces result from specific space particle flow interactions. Many "everyday" type forces; i.e. pushing/pulling matter are the result of proportioned directionalized gravity.*

**Hypothesis:** *Movement of matter is possible only because the space surrounding matter is consumed by the hole-complexes.*

**Hypothesis:** *Energy replaces the space being consumed in the direction of movement allowing for or causing a change in the consumptive pattern.*

**Hypothesis:** *Energy is a volume of space squared per time squared. Energy results in a volume of space per time returning to the outer-verse from the inner-verse.*

**Hypothesis:** *Matter and energy are equal but opposite processes. Whereas matter results from the consumption of space from the outer-verse, returning space to the inner-verse; energy returns space to the outer-verse from the inner-verse. The difference between matter and energy is the direction of space particle flow; meaning the hole-complexes are disconnected from each other and the flow of space particles from the outer-verse to the inner-verse is reversed.*

**Hypothesis:** *Matter can be viewed as energy exiting the outer-verse to the inner-verse via Einstein's equation, $E = mc^2$: i.e. $E_{oo} = \varsigma c^2$ where $E_{oo}$ is the energy transferred out of the outer-verse.*

**Hypothesis:** *All detectable massless particles are reverse hole-complexes through which a net flow of space particles pass from the inner-verse to the outer-verse.*

**Hypothesis:** *All detectable massless particles can be considered to be delivery vehicles for a volume of space returning to the outer-verse.*

**Hypothesis:** *The return of space particles to the outer-verse is ongoing, with a specific volume of space returning to the outer-verse for every individual wavelength of electromagnetic energy.*

**Hypothesis:** *Charge is a directionalized moving consumption of a volume of space traveling at the speed of light.*

**Hypothesis:** *Entropy is a volume of space added to or subtracted from a system.*

**Hypothesis:** *Temperature is an accelerating volume of space, resulting from a net return of a volume of space to the outer-verse from the inner-verse.*

**Hypothesis:** *The universe can be described by a consistent system of mathematical equations.*



*Eric Mitchell Horn*

**Hypothesis:** Individual mathematical equations describe physical elements; i.e. building blocks, which can be combined to describe more complex physical elements.

**Hypothesis:** A system of physical elements and corresponding physical constants describes how the universe works.

**Hypothesis:**  All physical elements, i.e. scat/mass, energy, gravity, space, temperature, charge, etc. can be expressed in terms of powers of the speed of light and time.

**Hypothesis:** All physical constants can be expressed as ratios between various combinations of powers of $c$, the speed of light; $t_\varsigma$, the  scat-time; $t_V$, the volume-time associated with an object with mass, and other fundamental time constants and time ratios.

**Hypothesis:** The law of conservation of matter and energy translates to:  Hole-complexes can be rearranged, and "flipped," but the individual holes connecting the inner-verse to the outer-verse cannot be destroyed.

# References


1. Kaku, Michio (1999). *Introduction to Superstring and M-Theory* (2nd ed.). New York, USA: Springer-Verlag.

2. Feng, J (2010). *Dark Matter Candidates from Particle Physics and Methods of Detection*; Annual Review of Astronomy and Physics, 48(1) pp. 495-545, {https://doi.org/10.1146/annurev-astro-082708-101659}

3. Heisenberg, L; Bartelmann, M.; Brandenberger, R.; Refregiet, A.; *Dark Energy in the Swampland*; Phys. Rev. D 98, 123502, 6 Dec 2018

4. Ulmshneider, P.; Vial, J.;(1997) Bocchialini, K.; Boumier, P. eds.; *Heating of Chromospheres and Coronae in Space Solar Physics*; Proceedings, Orsay, France; Springer, pp. 77-106; ISBN 978-3-540-64307-4

5. Malara, F.; Velli, M.;(2001); Pål Brekke; Bernhard Fleck; Joseph Gurman, eds. *Observations and Models of Coronal Heating*; Recent Insights into the Physics of the Sun and Heliosphere: Highlights from SOHO and other Space Missions, Proceedings of IAU Symposium 203. Astronomical Society of the Pacific. pp.456-466; ISBN 978-1-58381-069-9

6. White, H.; March, P.; Lawrence, J.; Vera, J.; Sylvester, A.; Brady, D.; Bailey, P.; (2017) *Measurement of Impulsive Thrust from a Closed Radio-Frequency Cavity in Vacuum*, Journal of Propulsion and Power, 33(4), pp. 830-841. {https://doi.org/10.2514/1.B36120}

7. Fixsen, D.; (2009). *The Temperature of the Cosmic Microwave Background*; The Astrophysical Journal, **707** (2): 916–920.

8. Atkins, P.; dePaula, J.; *Physical Chemistry*, 9th ed., Oxford University Press

9. Neal, V.; Neshyba, S.; Denner, W.; Notes: Limnology and Oceanography; 17(3) MAY 1972, pp 451-454.

10. Abraham, J.;Baringer, M.;Bindoff, N.;Boyer, T.; *A Review of Global Ocean Temperature Observations: Implications for Ocean Heat Content Estimates and Climate Change*; Reviews of Geophysics 51(3):450-483 August 2013

11. Diniz, F.; Ramos, A.; Rebello, E.; (2018) *Brazilian Climate Normals for 1981-2010. Pesq. agropec. bras.* [online]., 53(2) pp.131-143. Available from: <http://www.scielo.br/scielo.php? http://doi.org/10.1590/s0100-204x2018000200001.




*Eric Mitchell Horn*

12. Anzellini, S.; Dewaele, A.; Mezouar, M.; Loubeyre, P.;  Morard, G.;(2013) *"Melting of Iron at Earth's Inner Core Boundary Based on Fast X-ray Diffraction"*; Science, AAAS. **340** (6136): 464–466. {https//doi.org/10.1126/science.1233514}

13. Knuuttila, Tauno; (2000) "Nuclear Magnetism and Superconductivity in Rhodium"; Helsinki University of Technology; ISBN 951-22-5214-7 p 22.

# Glossary of Selected Terms and Symbols

This theory contains many new terms and symbols essential to the theory. Some symbols will have different meanings from convention. The variety of symbols employed in this theory may be confusing to first time readers. Reading these symbols backwards may help alleviate this confusion. For example, $t_{\varsigma Q_e}$ symbolizes the "electron charge scat-time" and $t_{\varphi B_{Q_e}}$ symbolizes the "electron charge magnetic flux time."

$a_g$- The symbol for gravitational acceleration with a $ct$ expression of $ct^{-1}$ and a value equal to $\dfrac{c}{t_{g'}} = 6.6741 \times 10^{-11}$m/s²

$A$- The symbol for the physical element of area with a $ct$ expression of $c^2t^2$

$A_a$ -The symbol for the physical element of accelerated area with a $ct$ expression of $c^3t$

$A_{a'}$ - The symbol for the subset of accelerating area; $A_{a'} = nc^2$ with a $ct$ expression $c^2$

$A_c$ - The symbol for the subset of the consumptive area of a an object with mass; includes only a time factor necessary to determine an area. $A_c$ , the consumptive area, is an area derived from a volume and has a $ct$ expression of $c^2t_V^2$.

$A_{\varphi B_{Q_e}}$ - The symbol for the electron charge magnetic flux area with $ct$ expression of $c^2t^2$ and a value equal to $c^2t^2_{\varphi B_{Q_e}} = 3.0201 \times 10^{-10}$m²

$A_\varsigma$- The symbol for the subset of scar with a $ct$ expression of $c^2t_\varsigma^2$; a scat-area of an object with mass, with a scat-time factor.

$A_\varsigma^2$ - The symbol for scar squared, with a $ct$ expression of $c^4t_\varsigma^4$ , which is equal to the scat of an object with mass

$A_{\varsigma P}$ - The symbol for the Planck scar, with a $ct$ expression of $c^2t^2$ and a  value equal to $c^2t^2_{\varsigma P} = 2.0677 \times 10^{-8}$m²

acar - The acronym for the physical element of accelerated area with a $ct$ expression of $c^3t$; an accelerated area is also an accelerating space.

acdi - The acronym for the physical element of accelerated distance with a $ct$ expression of $c^2$; an accelerated distance is also an accelerating area.

acrat - The acronym for physical element of accelerated ratio with a $ct$ expression of $ct^{-1}$; accelerating distance; an accelerated unit-less ratio (n) and a $ct$ equation
$$acrat = \frac{c}{t} = \frac{nc}{t} = \frac{c}{t'}, \text{ where } t' = \frac{n}{t} \text{ or just } t$$

acscat - The acronym for the physical element of accelerated scat with a $ct$ expression of $c^5t^3$; accelerated scat is also accelerating scop. Gravity and force are subsets.

acscop - The acronym for the physical element of accelerated scop with a $ct$ expression of $c^6t^4$; energy; accelerated scop is also accelerating space squared

acsess - The acronym for the physical element of accelerated sustained energy/space squared, symbolized $ACS$, with a $ct$ expression $c^7t^5$

$ACS_\varsigma$- The symbol for accelerated space squared due to an object with mass

$ACS_E$- The symbol for accelerated space squared due to energy

acsp - The acronym for the physical element of accelerated space with a $ct$ expression of $c^4t^2$; accelerated space is also accelerating scat; inertial mass and current are subsets.

$\alpha$ - The symbol for the fine structure constant with a value equal to $\dfrac{t_{R_K}}{2t_f}$

$c^2$ - The symbol for the speed of light squared; an accelerating area or an accelerated distance

$ct$ expression- A speed of light and time expression/equation for a physical element

$ct$ expression analysis- Similar to dimensional analysis, an analysis that relates general equations for physical elements such as energy, scat/mass, space, temperature, entropy, gravity, access, etc.

$C_{c_0}$ - The symbol for a quantum circle circumference equal to
$$2\pi ct_f = 4.9542 \times 10^{-43}\text{m}$$

$d_{a'}$ - The symbol for accelerating distance; traditional acceleration, $a$, is an accelerating distance, with a $ct$ expression of $ct^{-1}$

$d_{c0}$ - The symbol for a quantum circle diameter equal to $2ct_f = 1.5770 \times 10^{-43}$m

directionalized - A reference to a physical element with the property of direction

$E$- The symbol for the physical element of energy, with a $ct$ expression of $c^6 t^4$

$E_0$ - The symbol for quantum energy, with a value equal to $1.5764 \times 10^{-163}$m⁶/s²
$= 7.0810 \times 10^{-145}$J

$E_h$- The symbol for the Hartree energy constant

$E_P$ - The symbol for Planck energy, the energy associated with Planck's constant with

with a $ct$ expression of $c^6 t^4$ and a value equal to $c^6 t^4_{\varsigma P} = \dfrac{S^2_{Q_e}}{t^2_{R_K}} = 38.425$ m⁶/s²

$= 1.7260 \times 10^{20}$ J

$E_{io}$ - The symbol for "energy incoming outer-verse"; energy entering the outer-verse

$E_{oo}$ - The symbol for "energy outgoing outer-verse"; energy leaving the outer-verse

$E_S$ - The symbol for the energy of space, space energy

$e$ - The symbol for the electron

$e'$ - The symbol for the derived charge constant associated with the electron; with SI derived units of m⁴/s, a $ct$ expression of $c^4 t^3_\varsigma$ and a value equal to $c^4 t^3_{\varsigma Q_e}$. For purposes of disambiguation from the electron, the symbol $Q_e$ will generally be used to denote the value of electron charge.

$\varepsilon'_0$ - The symbol for the derived electric constant with a $ct$ expression of $ct$ and a value equal to $ct_f = 7.8848 \times 10^{-44}$m

fermionic- An adjective, relating to fermions, i.e. quarks and leptons which compose matter.

$F$ - The symbol for the physical element of force; "directionalized" gravity with a time element and a $ct$ expression of $c^5t^3$; a subset of acscat

$F_{g1kg}$ - The symbol for the gravitational force, in newton's, of a 1 kilogram mass

$F''_Q$ - The symbol for the force acting between two charges

$F''_{Q_e}$ - The symbol for the repulsive force acting between two electron charges one meter apart

$G'_0$ - The symbol for the derived conductance quantum with a $ct$ expression of $c^2t$

$G_F^{0'}$ - The symbol for the derived Fermi coupling constant with a $ct$ expression of $c^{-12}t^{-8}$

$G'_F$ - The symbol for the numerator of the derived Fermi coupling constant with a $ct$ expression of $c^9t^7$.

$G_{kN}$ - The symbol for Newton's gravitational constant

$g$ - The symbol for the physical element of gravity, gravitational force with a $ct$ expression of $c^5t^3$

$g'$ -The symbol for the gravitational force exerted by one object with mass with a $ct$ expression of $c^5t^3$

$g'_{obs}$ - The symbol for the observed or measured gravity of one object with mass, $\varsigma$ or $\varsigma'$, where $\varsigma' = \sqrt{\varsigma_1\varsigma_2}$

$g''$ - The symbol for the gravitational force of attraction acting between two objects with mass expressed in m⁴ with a $ct$ expression of $c^5t^3$

$g''_{Q_e}$ - The symbol for the gravitational force acting between two scat equivalents of electron charge one meter apart

$g_0$ - The symbol for hypothesized quantum gravity with $ct$ expression of $c^5t^3$ with a value equal to $1.1646 \times 10^{-190}$m⁵/s² $= 5.2313 \times 10^{-177}$N

$h'$ - The symbol for the derived Planck constant with units of m⁶/s and a $ct$ expression of $c^6t^5$ and a value equal to $\dfrac{S_{Q_e}^2}{t_{R_K}\psi} = \dfrac{c^6t_{\varsigma_{Q_e}}^6}{t_{R_K}\psi} = 1.4751 \times 10^{-52}$m⁶/s$\psi$

$h'_b$ - The symbol for the derived $h$ barr constant with units of m⁶/s and a $ct$ expression of $c^6t^5$ and a value equal to $\dfrac{c^6t_{\varsigma_{Q_e}}^6}{2\pi t_{R_K}} = 2.3477 \times 10^{-53}$m⁶/s

$H_0$ - The symbol for quantum magnetic field strength, the inverse of the derived magnetic constant with a $ct$ expression of $c^3t$ and a value equal to $c^3t_f = 7.0865 \times 10^{-27}$m³/s²

inner-verse - The part of the universe, symbolized $U_i$ , which is currently unobservable and will likely remain so even with improvements in technology

$I_{Q_e}$ - The symbol for the electron charge accelerated space with a $ct$ expression of $c^4t^2$ and a value of $c^4t_{\varphi_{B_{Q_e}}}^2 = 2.7143 \times 10^7$m⁴/s²

$k_{B'}$ - The symbol for the derived Boltzmann's constant with SI derived units of m³/electron or atom or molecule and a $ct$ expression of $c^3t^3$/electron or atom or molecule and a value equal to $c^3t_{k_B}^3 = 1.5812 \times 10^{-41}$m³

$k_{e''}$ - The symbol for the derived electric charge constant relating the force acting between two charges, with SI derived units of m⁻¹ and a $ct$ expression of $c^{-1}t^{-1}$ and a value equal to $\dfrac{1}{4\pi c t_f} = \dfrac{1}{9.9084 \times 10^{-43}}$m⁻¹ $= 1.0092 \times 10^{42}$m⁻¹

$k_{g'}$ - The symbol for the derived gravitational constant for one object with mass with SI derived units of m/s² and a $ct$ expression of $ct^{-1}$ and a value equal to $\dfrac{c}{t_{g'}} = 6.6741 \times 10^{-11}$m/s²

$k_{g''}$ - The symbol for the gravitational constant relating the force of attraction between two objects with mass (two rest masses) with a $ct$ expression of $c^{-1}t^{-3}$ and a value equal to $\dfrac{1}{ct_{g''}^3} = 2.9979 \times 10^8$ m⁻¹s⁻³

$k_{J'}$ - The symbol for the derived Josephson constant with a $ct$ expression of $c^{-2}t^{-2}$ and a value equal to $\dfrac{2}{c^2 t_{\varsigma P}^2} = 9.6726 \times 10^7$ m⁻²

$k_{tfg}$ - The symbol for the time ratio of the fundamental time constant to the gravitational time constant and a value equal to $\dfrac{t_f}{t_{g'}} = 5.8551 \times 10^{-71}$ unit-less

$k_{tkg\varsigma}$ - The symbol for the mass conversion constant relating an object with mass expressed in kilograms to that same object with scat expressed in m⁴; a unit-less value equal to $2.2262 \times 10^{-19}$

m - The symbol for the SI unit of measure for distance, the meter

$m$ - The traditional symbol for mass

$m_i$ - The symbol for inertial mass, a physical element with a $ct$ expression of $c^4 t^2$; a subset of acspace, (accelerated space) resulting in an accelerating mass.

$m_{m^4}$ - The symbol for a rest mass quantity expressed in m⁴ with a $ct$ expression of $c^4 t^4$

$m_{Q_e}$ - The symbol for the electron charge mass equivalent with a value equal to $c^4 t_{\varsigma Q_e}^4$

modi - An acronym for moving distance, a physical element with a $ct$ expression of $c^2 t$; voltage and conductance are subsets.

moar - An acronym for moving area, a physical element with a $ct$ expression of $c^3 t^2$; $\sqrt{E}$, $cA_\varsigma$ are subsets.

mova - An acronym for a moving volume-area, a moving area derived from a volume, with a $ct$ equation $mova = c^3 t_V^2$

moen - An acronym for moving energy with a $ct$ expression of $c^7 t^4$

moent - An acronym for moving energy through time; energy moved a distance; a subset of access with a $ct$ expression of $c^7t^5$

mosa - An acronym for a moving scat-area with a $ct$ equation $mosa = c^3t_\varsigma^2$

moscat - An acronym for moving scat; momentum with a $ct$ expression of $c^5t^4$

moscop - An acronym for moving scop, a physical element with a $ct$ expression of $c^6t^5$; Planck's constant is a subset.

mosp- An acronym for moving space, a physical element with a $ct$ expression of $c^4t^3$. Mosp includes two subsets, mov and most.

most- An acronym for moving space with a time element, a subset of mosp, with a $ct$ expression of $c^4t_\varsigma^3$; Electric charge is a subset of most.

mov - An acronym for moving volume, a subset of mosp, with a value equal to $c^4t_V^3$. Mov has no time element, other than $t_V$ which specifies a consumptive volume of space, $(V_c)$.

$mov_0$ - The symbol for quantum moving volume with a value equal to $3.8847 \times 10^{-199}$ m⁴/s

$\mu_0'$ - The symbol for the derived magnetic constant with a $ct$ expression of $c^{-3}t^{-1}$ and a value equal to $\dfrac{1}{c^3t_f} = 1.4111 \times 10^{26}$ s²/m³

$n$ - A unit-less number; a ratio of physical elements ( force, mass, volume, area, length, speed, time, etc.) which results in a unit-less number

$N_A$ - The symbol for Avogadro's number

outer-verse - The part of the universe, symbolized $U_o$ which can be observed or will likely be able to be observed with improvements in technology

$P$ - The symbol for pressure;  a subset of acar (accelerated area) with a $ct$ expression of $c^3t$

$p_{max}$- The symbol for maximum momentum, moving scat, a physical element with a $ct$ expression of $c^5t^4$

photonic- An adjective, relating to the photon

photonic cycle- An electromagnetic energy cycle in which a volume of space, equal to the volume of space consumed by the charge of the electron, is returned to the observable universe.

$\varphi_{B_{Q_e}}$- The symbol for the electron charge magnetic flux with a $ct$ expression of $c^2t^2$ and a value equal to $c^2t^2_{\varphi_{B_{Q_e}}} = 3.0201 \times 10^{-10} \text{m}^2$

$\varphi_{E_{Q_e}}$- The symbol for the electron charge electric flux with a $ct$ expression of $c^3t^2$ and a value equal to $c^3t^2_{\varphi_{B_{Q_e}}} = 0.090539 \text{ m}^3/\text{s}$

$\psi$ - The symbol for photonic cycle(s)

physical element - a mathematical representation of an essential component or "building block" that forms the framework of this theory

$Q$ - The symbol for charge with derived SI units of m⁴/s and a $ct$ expression of $c^4t^3$

$Q_e$ - The symbol for the derived electron charge quantity, $e'$, with a $ct$ expression of $c^4t^3$ and a value equal to $c^4t^3_{\varsigma_{Q_e}} = -7.1338 \times 10^{-45} \text{m}^4/\text{s}$

$Q_{q_b}$ - The symbol for the bottom quark charge with a $ct$ expression of $c^4t^3$ and a value equal to $c^4t^3_{\varsigma_{Q_{q_b}}} = -2.3779 \times 10^{-45} \text{m}^4/\text{s}$

$Q_{q_c}$ - The symbol for the charm quark charge with a $ct$ expression of $c^4t^3$ and a value equal to $c^4t^3_{\varsigma_{Q_{q_c}}} = +4.7559 \times 10^{-45} \text{m}^4/\text{s}$

$Q_{q_d}$ - The symbol for the down quark charge with a $ct$ expression of $c^4t^3$ and a value equal to $c^4t^3_{\varsigma_{Q_{q_d}}} = -2.3779 \times 10^{-45} \text{m}^4/\text{s}$

$Q_{q_s}$ - The symbol for the strange quark charge with a ct expression of $c^4t^3$ and a value equal to $c^4t^3_{\varsigma_{Q_{q_s}}} = -2.3779 \times 10^{-45} \text{m}^4/\text{s}$

$Q_{q_t}$ - The symbol for the top quark charge with a $ct$ expression of $c^4t^3$ and a value equal to $c^4t^3_{\varsigma_{Q_{q_t}}} = +4.7559 \times 10^{-45} \text{m}^4/\text{s}$

$Q_{q_u}$ - The symbol for the up quark charge with a $ct$ expression of $c^4t^3$ and a value equal to $c^4t^3_{\varsigma_{Q_{q_u}}} = +4.7559 \times 10^{-45} \text{m}^4/\text{s}$

quantum circle - A theoretical quantum circle with a radius equal to

$$\varepsilon_0' = ct_f = 7.8848 \times 10^{-44}\text{m}$$

$r_{c0}$ - The symbol for the radius of a quantum circle, $r_{c0} = \varepsilon_0'$

$R$ - The symbol for resistance with a $ct$ expression of $c^{-2}t^{-1}$

$R'$ - The symbol for the derived gas constant, a volume of space per mole with a $ct$ expression of $\dfrac{c^3t^3}{mol}$ and a value equal to $9.5226 \times 10^{-18}$m³/mol

$R_K'$ - The symbol for the derived von Klitzing constant with a $ct$ expression of $c^{-2}t^{-1}$ and a value equal to $\dfrac{1}{c^2t_{R_K}} = 2.8985 \times 10^{36}$s/m²

$R_K^{-1}$ - The symbol for the inverse of the derived von Klitzing constant with a $ct$ expression of $c^2t_{R_K}$ and a value equal to $3.4501 \times 10^{-37}$m²/s.

scat - A symbolized name for space multiplied by the speed of light and time; a physical element that quantifies matter; an equation which specifies rest mass expressed in m⁴; with a $ct$ expression of $c^4t^4$

$\varsigma$ - The symbol for scat

$\varsigma_0$ - The symbol for quantum scat with a value equal to $1.7540 \times 10^{-180}$m⁴ $= 7.8788 \times 10^{-162}$kg

$\varsigma_e$ - The symbol for the scat of the electron

$\varsigma_p$ - The symbol for the scat of the proton

$\varsigma_P$ - The symbol for the scat equivalent of Planck's constant with a $ct$ expression of $c^4t^4$ and a value equal to $c^4t_{\varsigma_P}^4 = A_{\varsigma_P}^2 = 4.2754 \times 10^{-16}$m⁴ $= 1{,}920.5$ kg

$\varsigma_{Q_e}$ - The symbol for the electron charge scat equivalent with a value equal to $c^4t_{\varsigma_{Q_e}}^4 = 6.8444 \times 10^{-71}$m⁴ $= 3.0745 \times^{-52}$ kg

$\varsigma_{1kg}$ - The symbol for the scat of one kilogram with a $ct$ expression of $c^4t^4$ and a value equal to $c^4t^4_{\varsigma_{1kg}} = 2.2262 \times 10^{-19}$m⁴

scop- An acronym for sustained consumptive pattern, symbolized $SCP$, a physical element quantitatively equal to scat/mass moved a distance, with a $ct$ expression of $c^5t^5$

spart - An acronym for space particle

$S$ - The symbol for space or a volume of space and the symbol for entropy: a physical element with a $ct$ expression of $c^3t^3$

$S_{a'}$ - The symbol for accelerating space; interchangeable with $V_{a'}$, accelerating volume, with a $ct$ expression of $c^3t$

scar - An acronym for scat-area, symbolized $A_\varsigma$, with a $ct$ equation $scar = A_\varsigma = c^2t^2_\varsigma$

$S_{con}$ - The symbol for conductance, (disambiguation from $S$, the symbol for space)

$S_0$ - The symbol for the hypothesized volume of the space particle with a $ct$ expression of $c^3t^3_{S_0}$ and a value equal to $1.2958 \times 10^{-207}$m³

$SE_E$ - The symbol for the space equivalent of energy with a $ct$ expression of $c^3t^3$

$SE_\varsigma$- The symbol for the space equivalent of scat with a $ct$ expression of $c^3t^3$

sess- An acronym for sustained energy space squared, a physical element with a $ct$ expression of $c^6t^6$

$spin_f$ - The symbol for the quantum spin of fermions, the intrinsic fermionic angular momentum, with a $ct$ expression of $c^6t^5$ and a value equal to

$$\frac{h_b}{2} = \frac{S^2_{Q_e}}{4\pi t_{R_K}} = \frac{S^2_{Q_e}}{t_{G_0}\psi} = 1.1739 \times 10^{-53}\text{m⁶/s}$$

$S_{Q_e}$ - The symbol for the electron charge space with a value equal to $c^3t^3_{\varsigma_{Q_e}}$. Note that the symbol for the electron charge scat-volume refers to the same thing; i.e.

$$V_{\varsigma_{Q_e}} = S_{Q_e} = c^3t^3_{\varsigma_{Q_e}} = 2.3796 \times 10^{-53}\text{m³}$$

$S^2$ - The symbol for sustained energy space squared

$S^2_{Q_e}$ - The symbol for the electron charge space squared with a value equal to $c^6 t^6_{\varsigma_{Q_e}} = 5.6625 \times 10^{-106} \text{m}^6$

$t$ - The symbol for time or a time-length

$t_{E_h}$ - The symbol for the scat-time of Hartree energy constant

$t_f$ - The symbol for the time-length of the fundamental time constant with a value equal to $2.6301 \times 10^{-52}\text{s}$

$t_{g'}$ - The symbol for the time-length of the gravitational time constant relating gravity to one object with mass with a value equal to $4.4920 \times 10^{18}\text{s}$

$t_{g''}$ - The symbol for the time-length of the gravitational time constant relating two objects with mass with a value equal to $2.2325 \times 10^{-6}\text{s}$

$t_{G_0}$ - The symbol for the time-length of the conductance quantum with a value equal to $7.6774 \times 10^{-54}\text{s}$

$t_{G_F}$ - The symbol for the time-length of the of the numerator of the derived Fermi-coupling constant

$t_{k_B}$ - The symbol for the scat-time of Boltzmann's constant; $t_{k_B} = t_{\varsigma k_B} = 8.3722 \times 10^{-23}\text{s}$

$t_{k_J}$ - The symbol for the time-length of Josephson's constant with a value of $\dfrac{t_{\varsigma P}}{\sqrt{2}} = 3.3916 \times 10^{-13}\text{s}$

$t_{R_K}$ - The symbol for the volume-time of the von Klitzing constant; $t_{R_K} = t_{V_{R_K}} = t_0 = 3.8387 \times 10^{-54}\text{s}$

$t_P$ - The symbol for Planck time

$t_p$ - The symbol for a pressure time-length

$t_{\varphi B_{Qe}}$ - The symbol for the electron charge magnetic flux time with a value equal to $5.7968 \times 10^{-14}$s

$t_{\varphi E_{Qe}}$ - The symbol for the electron charge electric flux time with a value equal to
$$t_{\varphi E_{Qe}} = t_{\varphi B_{Qe}} = 5.7968 \times 10^{-14}\text{s}$$

$t_\psi$ - The symbol for the photonic cycle time-length

$t_{S_0}$ - The symbol for the volume-time of the space particle

$t_\varsigma$ - The symbol for a scat-time

$t_{\varsigma p}$ - The symbol for the proton scat-time

$t_{\varsigma P}$ - The symbol for Planck scat-time; the scat-time equivalent of Planck's constant with a value equal to $4.7965 \times 10^{-13}$s

$t_{\varsigma Qe}$ - The symbol for the electron charge scat-time with a value equal to $9.5943 \times 10^{-27}$s

$t_{\varsigma Qe}^3$ - The symbol for the electron charge scat-time cubed with a value equal to $8.8316 \times 10^{-79}$s³

$t_T$ - The symbol for a temperature time-length

$t_V$ - The symbol for a volume-time

$t_{V_{Qe}}$ - The symbol for the electron charge volume-time with a value equal to $1.2356 \times 10^{-41}$s

$T$ - The symbol for temperature with a $ct$ expression of $c^3 t$

$T_0$ - The symbol for quantum temperature with a $ct$ expression of $c^3t$ and a value equal to $c^3t_{S_0} = 1.0343 \times^{-28}$ m³/s² $= 5.3210 \times 10^{-28}$K

$T_P$ - The symbol for Planck temperature with a $ct$ expression of $c^3t$ and a value equal to $c^3t_{\varsigma P} = 1.2924 \times 10^{13}$m³/s² $= 6.6487 \times 10^{13}$K

$T_S$ - The symbol for the temperature of space; space temperature

$T_u$ - The symbol for the Tesla unit of measure (unit-less) with a value equal to $5 \times 10^6$

$T_{loss_g}$- The symbol for the gravitational loss of temperature

$U_i$ - The symbol for the inner-verse; the part of the universe which cannot be observed by currently available technology and is likely to remain unobservable even with improvements in technology

$U_o$ - The symbol for the outer-verse; the part of the universe which can be observed, commonly the "universe"

$V_a$ - The symbol for an accelerated volume; interchangeable with $S_a$, the symbol for accelerated space with a $ct$ expression of $c^4t^2$

$V_{a_\varsigma'}$- The symbol for an accelerating volume of space due to an object with mass, a physical element with a $ct$ expression $c^3t$, An accelerated area results in accelerating space, with an SI derived unit of m³/s² ; $V_{a_\varsigma'}$ is accelerating space towards a hole-complex in space.

$V_{a_E'}$ - The symbol for an accelerating volume of space due to energy; temperature and pressure are subsets, also with a $ct$ expression of $c^3t$; $V_{a_E'}$ is the accelerating volume of space away from a hole-complex in space.

$V_c$ - The symbol for the consumptive volume of an object with mass or the consumptive volume of a mass equivalent of energy: the volume of space which is traveling at the speed of light immediately before passing from the outer-verse to the inner-verse with a $ct$ expression of $c^3t_V^3$.

$V_{c_{H_0}}$ - The symbol for the quantum magnetic field strength consumptive volume equivalent

$V_{ck_B}$ - The symbol for the Boltzmann constant consumptive volume equivalent

$V_{ck_e}$ - The symbol for the Coulomb constant consumptive volume equivalent

$V_{cP}$ - The symbol for the Planck constant consumptive volume equivalent

$V_{cR_K}$ - The symbol for the von Klitzing constant consumptive volume equivalent

voar - An acronym for a volume-area, (see next,) with a $ct$ equation
$voar = A_V = c^2 t_V^2$ . Voar does not have a time element other than to describe an area derived from a volume.

volume-area - An area derived from a volume, with a $ct$ expression of $c^2 t_V^2$ . Volume-area and voar refer to the same subset of area. The term "volume-area" may lead to confusion,  and therefore will generally not be used other than to illustrate the derivation of the term voar.

$V_S$ - The symbol for a specific volume of space (may be interchanged with $S$) with a $ct$ expression of $c^3 t^3$

$V_{S_\psi}$ - The photonic cycle space volume, the volume of space returned per photonic cycle with a $ct$ expression of $c^3 t^3$ and a value equal to $S_{Q_e} = V_{\varsigma Q_e} = c^3 t_{\varsigma Q_e}^3$

$V_\varsigma$ - The symbol for scat-volume, a volume of space consumed by an object with mass in one scat-time with a $ct$ expression of $c^3 t_\varsigma^3$

$V_{\varsigma Q_e}$ - The symbol for the electron charge scat-volume with a $ct$ expression of $c^3 t^3$ and a value equal to $c^3 t_{\varsigma Q_e}^3$ .The notation, $S_{Q_e}$ , for electron charge space refers to the same thing; i.e $V_{\varsigma Q_e} = S_{Q_e} = c^3 t_{\varsigma Q_e}^3$

$V_{ltg}$ - The symbol for voltage (disambiguation from volume); a moving distance, a subset of modi with a $ct$ expression of $c^2 t$

$\nu$ - The symbol for the photonic cycle frequency

$Z'_0$ - The symbol for the derived characteristic impedance of a vacuum constant with a $ct$ expression of $c^{-2}t^{-1}$ and a value equal to $\dfrac{1}{c^2 t_f} = 4.2304 \times 10^{34}\text{s/m}^2$

$$Acsess = c^7 t^5 \qquad\qquad Sess = c^6 t^6$$

$$Moen = c^7 t^4 \qquad Moscop = c^6 t^5 \qquad Scopt = c^5 t^6$$

$$Energy = c^6 t^4 \qquad Scop = c^5 t^5$$

$$Moacscat = c^6 t^3 \qquad Momentum = c^5 t^4 \qquad Spatdi = c^4 t^5$$

$$Acscat = c^5 t^3 \qquad Scat = c^4 t^4$$

$$Moacsp = c^5 t^2 \qquad Mosp = c^4 t^3 \qquad Spat = c^3 t^4$$

$$Acsp = c^4 t^2 \qquad Space = c^3 t^3$$

$$Moacar = c^4 t \qquad Moar = c^3 t^2 \qquad Arti = c^2 t^3$$

$$Acar = c^3 t \qquad Area = c^2 t^2$$

$$Moacdi = c^3 \qquad Modi = c^2 t \qquad Motisq = c t^2$$

$$Acdi = c^2 \qquad Distance = ct$$

$$Moacrat = c^2 t^{-1} \qquad Motion = speed = c \qquad Time = t$$

$$Acrat = c t^{-1} \qquad Ratio = number = n \qquad Inmoti = c^{-1} t$$

$$Mointisq = c t^{-2} \qquad Inti = t^{-1} \qquad Inmo = c^{-1}$$



Eric Mitchell Horn

$$Intisq = t^{-2} \qquad Indi = c^{-1}t^{-1} \qquad Inacdi = c^{-2}$$

$$Inticu = t^{-3} \qquad Inmointisq = c^{-1}t^{-2} \qquad Inmodi = c^{-2}t^{-1}$$

$$Inmointicu = c^{-1}t^{-3} \qquad Inar = c^{-2}t^{-2} \qquad Inacar = c^{-3}t^{-1}$$

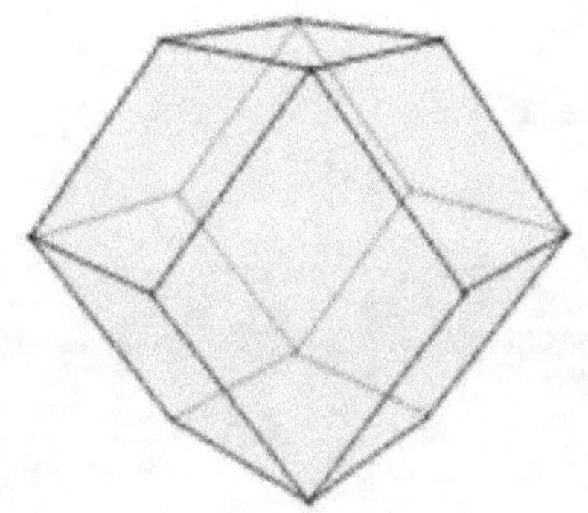

$$k_{g''} = \frac{g''_{Q_e}}{F''_{Q_e} 4\pi c\, t_f t^2_{\varsigma Q_e}} = 2.9979 \times 10^8 \text{m}^{-1}\text{s}^{-2}$$

$$T_{loss_g} = c^3 t_\varsigma$$

$$Q_e = A_{\varsigma P} R_K^{-1} = c^2 t^2_{\varsigma P} * c * c t_{R_K} = c^2 t^2_{\varphi B_{Q_e}} * c * c t_f = A_{\varphi B_{Q_e}} Z_0^{-1}$$

$$t_f = \frac{t_{R_K}}{2\alpha} = \frac{\epsilon'_0}{c} = \frac{1}{c^2 Z'_0} = \frac{1}{4\pi c k'_e} = \frac{1}{c^3 \mu'_0} = \frac{H_0}{c^3}$$

$$t_{R_K} = \frac{1}{c^2 R'_K}$$

$$E = h'f = \frac{S^2_{Q_e}}{t_{R_K} t_\psi} = (\frac{V_S}{t})^2 = T^2 t^2 = ST = c^6 t^4$$

$$E_P = \frac{S^2_{Q_e}}{t^2_{R_K}}$$

$$\alpha = \frac{t_{R_K}}{2 t_f}$$

$$restmass = V_c * c * t = Scat = c^4 t^4_\varsigma$$

$$g_{base} = g' = c^2 V_c$$

$$g_{obs} = g_{base} * (1 - \frac{T}{T_{\varsigma P}})$$

$$g'' = k_{g''} \frac{\varsigma_1 \varsigma_2}{c^2 t^2_d}$$

$$T_S = \frac{E_S}{V_S}$$

$$S^2 = h\nu t^2 = E t^2$$

$$|SE_E| = |\sqrt{h\nu t^2}| = |t\sqrt{E}| = |ct A_\varsigma| = |SE_\varsigma|$$

$$t_{R_K} = \frac{1}{c^2 R'_K}$$

$$g_0 = c^2 S_0 = c^5 t^3_{R_K}$$



Eric Mitchell Horn

$$k_{e''} = \frac{1}{ct_{k_e}} = \frac{1}{4\pi c t_f} = \frac{\varphi_{B_{Q_e}}}{4\pi V_{\varsigma_{Q_e}}}$$

$$Z_0' = \frac{1}{c^2 t_f} = 2\alpha R_K'$$

$$R_K' = \frac{1}{c^2 t_{R_K}} = \frac{1}{2\alpha c^2 t_f}$$

$$G_0'^{-1} = \frac{1}{c^2 t_{G_0}} = \frac{1}{4\alpha c^2 t_f}$$

$$k_{g''} = \frac{g_{Q_e}''}{F_{Q_e}'' c t_{k_e} t_{\varsigma_{Q_e}}^2} = 2.9979 \times 10^8 \text{m}^{-1}\text{s}^{-2}$$

$$k_{J'} = \frac{4\alpha t_f}{c^2 t_{\varsigma_{Q_e}}^3} = \frac{2}{c^2 t_{\varsigma_P}^2}$$

$$\mu_0' = \frac{1}{c^3 t_f}$$

$$T = c^3 t_\varsigma$$

$$k_{g''} = \frac{g_{Q_e}''}{F_{Q_e}'' 4\pi c t_f t_{\varsigma_{Q_e}}^2} = 2.9979 \times 10^8 \text{m}^{-1}\text{s}^{-2} \qquad g_{base} = g' = c^2 V_c \qquad restmass = V_c * c * t_{g'} = Scat = c^4 t_\varsigma^4$$

$$E = h'f = \frac{S_{Q_e}^2}{t_{R_K} t_\psi} = (\frac{V_S}{t})^2 = T^2 t^2 = ST = c^6 t^4 \qquad t_f = \frac{t_{R_K}}{2\alpha} = \frac{\epsilon_0'}{c} = \frac{1}{c^2 Z_0'} = \frac{1}{4\pi c k_{e''}} = \frac{1}{c^3 \mu_0'} = \frac{H_0}{c^3}$$

Errata

The following text corrects some verbiage relating to quantum constants:

Page 96 regarding Planck temperature should read: "...Boltzmann's constant is not a true smallest quantum constant, since a volt is not the smallest quantum moving distance..."

The following text corrects and adds to the gravitational temperature loss equation, principally noted on page 156, but also noted on pages 13, 194, and 228.

Page 156; Space Particle Theory (SPT) gravitational temperature loss correction:

Temperature is an accelerated area; (an accelerated area yields an accelerating volume of space.)

The observable (measurable) gravitational loss of temperature $(T_{loss/kg})$ is given directly by Newton's gravitational constant:

$$G_{kN} = 6.6741 \times 10^{-11} \text{m}^3/\text{kg-s}^2$$

Note that this equals 6.6741 x 10⁻¹¹m³/s²/kg

Per theory: 1 K = 5.1446 m³/s²

The observable (predicted) loss of temperature per 1 kg is therefore (will be):

$$6.6741 \times 10^{-11} \text{m}^3/\text{s}^2/\text{kg} \ \text{X} \ 5.1446 \ \text{K/m}^3/\text{s}^2 = 3.4335 \times 10^{-10} \ \text{K/kg}$$

The temperature equivalent of 1 kg is given by:

$$c^3 t_{\varsigma 1kg} = c^3 * 7.2456 \times 10^{-14} = 1.0044 \times 10^{13} \text{K}$$

As an analogous comparison:

An observable (measurable) mass is 1 kg. However, $m = \dfrac{E}{c^2}$. This is a mass equivalent of energy, i.e.:

| **_Observable_** | vs. | **_Equivalent_** |
|---|---|---|
| 1 kg | | $m = \dfrac{E}{c^2}$ |
| 3.4335 x 10⁻¹⁰K/kg | | $T = c^3 t_{\varsigma 1kg}$ |

The above is testable in an experiment.

Consider Newton's gravitational equation:

$$F_g = G_{kN} * \frac{m_1 * m_2}{d^2} \quad \text{with units of } G_{kN} = \frac{m^3}{kg \bullet s^2}$$

Applying $ct$ analysis to Newton's gravitational equation yields:

$$F = \frac{c^3 t^3}{c^4 t^4 * t^2} * \frac{c^4 t^4 * c^4 t^4}{c^2 t^2} = c^5 t^3$$

For "something" to be a force, it must obey Newton's second law. Applying $ct$ analysis to Newton's second law:

$$F = ma; \quad m = c^4 t^4 \quad a = \frac{c}{t} \quad (m = restmass)$$

$$F = c^4 t^4 * \frac{c}{t} = c^5 t^3$$

Newton's equation is shown (per theory) to relate two consumptive areas ($A = c^2 t^2$), not two masses per se, with $m'$ being the mass at the center of mass between the bodies:

$$m' = \sqrt{m_1 * m_2} = \sqrt{c^4 t_1^4 * c^4 t_2^4}$$

$$F_{g'} = g' = c^3 t * \frac{c^2 t_1^2 * c^2 t_2^2}{c^2 t^2} = c^5 t^3$$

Consider that temperature is a subset of the physical element of acar. An accelerated area ($c^3 t$; temperature, pressure, magnetic field strength) applied to [over] an area results in force:

$$c^3 t * c^2 t^2 = c^5 t^3 = F$$

10 kilograms is a number (e.g.;10) multiplied by 1 kilogram. 10 meters squared is a number (e.g.;10) multiplied by 1 meter squared. Newton's gravitational law can therefore be written as:

$$g'' = \frac{c^3 t}{kg} * \frac{n_1 kg * n_2 kg}{n_3 c^2 t_{1m}^2} \qquad \text{where } 1kg = c^4 t_{\varsigma 1kg}^4$$

Note that the kilogram unit can cancel, yielding the gravitational equation:

$$g'' = \frac{n_1 n_2}{n_3} * c^3 t * \frac{c^4 t_{\varsigma 1kg}^4}{c^2 t_{1m}^2}$$

But, this can be interpreted as representing an equation for the gravity of one mass at the center of gravity of the two mass system. Combining "$n$'s" yields:

$$g' = n c^3 t * \frac{c^4 t_{\varsigma 1kg}^4}{c^2 t_{1m}^2}$$

Per theory:

$$g' = c^2 V_c = c^5 t_V^3$$

$$c^4 t_{\varsigma1kg}^4 = V_{c1kg} * c * t_{g'} = c^3 t_{V1kg}^3 * c * t_{g'}$$

Let $n_1 = n_2 = n_3 = 1$, then:

$$g' = c^3 t * \frac{c^4 t_{V1kg}^3 * t_{g'}}{c^2 t_{1m}^2}$$

Determining the $t$ in the $c^3 t$ term:

$$g' = c^3 \frac{t_{1m}^2}{t_{g'}} * \frac{c^2 t_{V1kg}^3 * t_{g'}}{t_{1m}^2} = c^5 t_{V1kg}^3$$

$$t_{1m} = 3.3356 \times 10^{-9}\text{s} \,, \, t_{g'} = 4.4919 \times 10^{18}\text{s}$$

$$t = \frac{t_{1m}^2}{t_{g'}} = 2.4770 \times 10^{-36}\text{s}$$

$$c^3 t = 6.6741 \times 10^{-11}\text{m³/s²}$$

The area to [over] which this accelerated area is applied:

$$t_{\varsigma1kg} = 7.2456 \times 10^{-14}\text{s}$$

$$A_{g'} = n c^2 t^2 = n \frac{c^2 t_{\varsigma1kg}^4}{t_{1m}^2} = n c^2 (2.4770 \times 10^{-36}\text{s²}) = n * 2.2262 \times 10^{-19}\text{m}^2$$

$$g' = n * 6.6741 \times 10^{-11}\text{m³/s²} * 2.2262 \times 10^{-19}\text{m}^2$$

$$g' = n * 1.4858 \times 10^{-29}\text{m⁵/s²}$$

This confirms the equation for a gravitational force of a scat/mass when the mass is expressed as $n$ kilograms (see page 195 of text.)

The above discussion shows that there exists a temperature ($c^3 t$) loss due to gravity. It is that temperature loss that results in gravity being a "pull" rather than a "push." Gravity results from a consumption of space, ($g' = c^2 V_c$.)



Eric Mitchell Horn